SpringerBriefs in Biochemistry and Molecular Biology

For further volumes:
http://www.springer.com/series/10196

Lucy W. Barrett · Sue Fletcher
Steve D. Wilton

Untranslated Gene Regions and Other Non-coding Elements

Regulation of Eukaryotic Gene Expression

 Springer

Lucy W. Barrett
Sue Fletcher
Steve D. Wilton
Centre for Neuromuscular
  and Neurological Disorders (CNND)
The University of Western Australia
Crawley, WA
Australia

ISSN 2211-9353   ISSN 2211-9361 (electronic)
ISBN 978-3-0348-0678-7   ISBN 978-3-0348-0679-4 (eBook)
DOI 10.1007/978-3-0348-0679-4
Springer Basel Heidelberg New York Dordrecht London

Library of Congress Control Number: 2013940950

Printed on acid-free paper

Springer is part of Springer Science+Business Media (www.springer.com)

Preface

The completion of the human genome project in 2003 estimated the number of human genes to be between 20,000 and 25,000. It was assumed that humans, being highly complex organisms, would have many more genes than less complex organisms. However, *Caenorhabditiselegans* (roundworm) is estimated to have around 20,000 genes, and the number of mice genes is also in the same range as humans. This revelation meant that organism complexity could not be the result of a higher number of genes. Although there was no correlation between complexity and the number of genes, there was a clear correlation with the relative amount of non-coding sequences in the genome. In humans, only around 1.5 % of the genome is protein-coding, while the rest consists of introns, regulatory sequences and non-coding RNA. In the 10 years since the completion of the human genome project, research has rapidly progressed and we are now beginning to understand the importance of non-coding sequences. This book aims to summarise current knowledge about the non-coding regions of the eukaryotic genome and the roles they play in gene regulation and expression.

Lucy W. Barrett
Sue Fletcher
Steve D. Wilton

Contents

Untranslated Gene Regions and Other Non-coding Elements

Regulation of Eukaryotic Gene Expression

Abstract There is now compelling evidence that the complexity of higher organisms correlates with the relative amount of non-coding RNA rather than the number of protein-coding genes. Previously dismissed as "junk DNA", it is the non-coding regions of the genome that are responsible for regulation, facilitating complex temporal and spatial gene expression through the combinatorial effect of numerous mechanisms and interactions, working together to fine-tune gene expression. The major regions involved in regulation of a particular gene are the 5′ and 3′ untranslated regions and introns. In addition, pervasive transcription of complex genomes produces a variety of non-coding transcripts that interact with these regions and contribute to gene regulation. This review discusses recent insights into the regulatory roles of the untranslated gene regions and non-coding RNAs in the control of complex gene expression, as well as the implications of this in terms of organism complexity and evolution.

Keywords Regulation · Expression · Non-coding · Untranslated · RNA · Control

1 Introduction

Over the last decade it has become increasingly apparent that regulation of gene expression in higher eukaryotes is a complex and tightly regulated process involving many different factors and levels of control. For a given gene, the untranslated gene regions, including the 5′ and 3′ untranslated regions (UTRs), and introns are the major regions involved in the regulation of expression (Fig. 1). Despite being dismissed as "junk" DNA for many years, intergenic regions have also been found to contribute to control of gene expression, and evidence of pervasive transcription throughout the genome (Carninci et al. 2005; Cheng et al. 2005; Birney et al. 2007), both sense and antisense (He et al. 2008), implicates a role for all regions of the genome, which makes sense in terms of evolution, as it is

L. W. Barrett et al., *Untranslated Gene Regions and Other Non-coding Elements*,
SpringerBriefs in Biochemistry and Molecular Biology,
DOI: 10.1007/978-3-0348-0679-4_1, © The Author(s) 2013

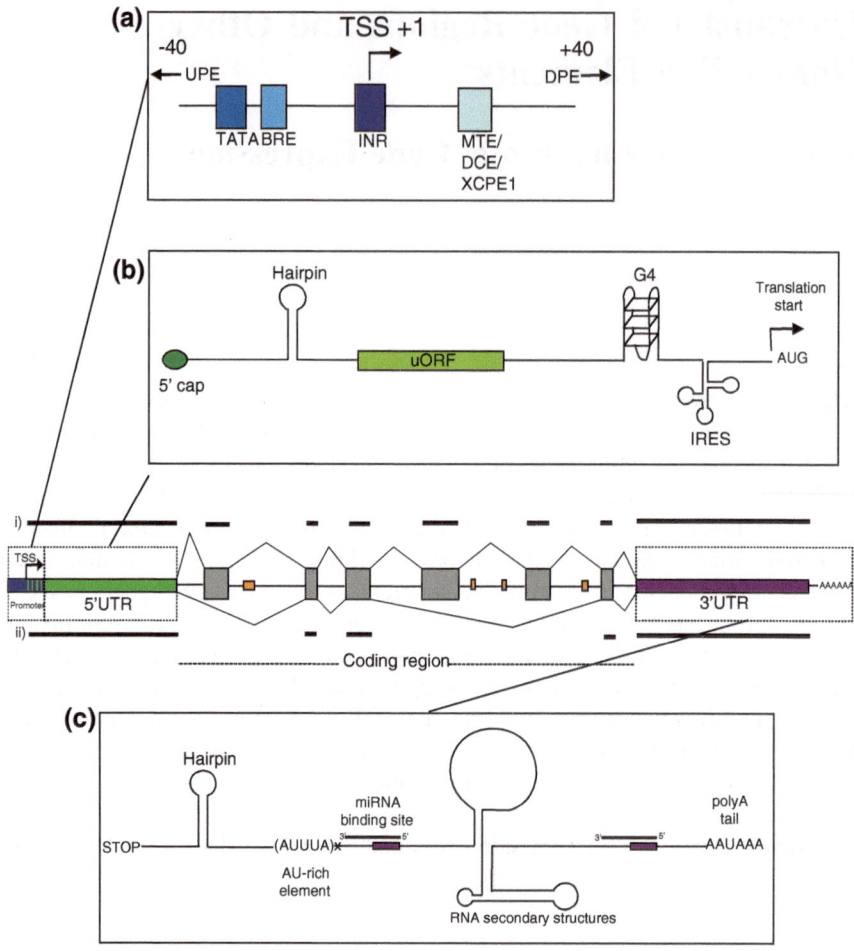

Fig. 1 Regulatory elements within the noncoding gene regions. The centre image shows a typical gene, with exons indicated in *grey*. The orange rectangles indicate intronic enhancer elements. The *black bars* indicate (i) regions included in the full-length transcript following splicing, and (ii) alternatively spliced transcript **a** Promoter region regulatory elements (adapted from Smale and Kadonaga 2003). Upstream and downstream promoter elements situated outside of the core promoter region are indicated by the *arrows*. **b** Regulatory elements in the 5'UTR. **c** Regulatory elements in the 3'UTR

expected that non-functional regions would be removed over time via natural selection. Accumulated evidence indicates that the complexity of higher organisms, which correlates with an increase in the size of non-coding regions, arises from an increase in the number and complexity of regulatory pathways (Levine and Tjian 2003), and that it is variation within these non-coding sequences that produces phenotypic variation between both individuals and species (Mattick 2001). This review will collate current knowledge concerning the role of untranslated gene regions, non-coding RNAs and other non-coding elements in the control of complex gene expression, with the aim of emphasising the complex mechanisms and interactions involved in precise gene control.

2 Promoter

The eukaryotic promoter is a regulatory region of DNA located upstream of a gene that binds Transcription factor II D (TFIID) and allows the subsequent coordination of components of the transcription initiation complex, facilitating recruitment of RNA polymerase II and initiation of transcription (Smale and Kadonaga 2003; Juven-Gershon et al. 2008). The core promoter generally spans ~80 bp around the transcription start site (TSS), and in mammals, can be separated into two distinct classes: conserved TATA-box enriched promoters that initiate at a single TSS (focused promoters), and variable CpG-rich promoters containing multiple TSS (dispersed promoters) (Carninci et al. 2006). The latter class is enriched in vertebrates, and expression from these promoters involves the combinatorial effects from a multitude of binding motifs within the promoter region. Some of the major elements involved in regulation by these complex promoters are enhancers, including upstream and downstream promoter elements (UPE and DPEs) that contain transcription factor binding sites and may act independently or synergistically with the core promoter to facilitate transcription initiation. Also commonly found in complex promoters are B-recognition elements (BRE), which are TFIIB recognition elements seven nucleotides in length that aid RNA polymerase II binding, and Initiator elements (INR), motifs that encompass the TSS and can act independently of, and synergistically with, TATA-box promoters via binding of TFIID (for a comprehensive review and details of each element refer to (Juven-Gershon et al. 2008; Smale and Kadonaga 2003). Other elements include insulators, activators, repressors, and some rarer, more recently discovered elements such as the motif ten element (MTE), downstream core element (DCE) and the X-core promoter element 1 (XCPE1), all of which act selectively with other elements to contribute to promoter activity (Fig. 1a) (Juven-Gershon et al. 2008). Notably, none of the core promoter elements identified thus far is ubiquitous or universally required for transcription (Yang et al. 2007), which is indicative of the complex and variable nature of promoters in eukaryotic genomes. The combination of multiple elements increases the potential for differential expression, and is influenced by the relative concentrations of interacting factors. In addition to core elements within the ~80 bp promoter region, identification of general functional regions using deletion analyses in multiple genes implicated the sequence lying −300 to −50 nucleotides upstream of the TSS as generally having a positive effect on promoter activity, while elements that negatively affected promoter activity were located −1000 to −500 nucleotides upstream of the TSS for 55 % of the genes tested (Cooper et al. 2006). Evidently it is not just the sequence in the immediate vicinity of the TSS that can influence promoter activity.

2.1 Types of Promoter

2.1.1 Focused Promoters

The TATA box is a well-characterised promoter motif with the consensus sequence TATAA. It is ancient, has been highly conserved throughout the evolution of eukaryotes, and was the first eukaryotic promoter element to be identified (Lifton et al. 1978). Located ~30 bases upstream of the TSS, the TATA box is bound by the TATA-box binding protein (TPB) (Burley and Roeder 1996) that induces a bend in the DNA, facilitating assembly of RNA polymerase II (PolII) and the general transcription factors TFIIB and TFIIA to initiate transcription (Patikoglou et al. 1999). The INR, located at +1 in all mammalian species, with the consensus YYANWYY also plays an important role in transcriptional control from TATA box promoters, through the binding of TFIID at the TSS (Yang et al. 2007). In addition to promoters controlled synergistically by both a TATA box and an INR, some lack an INR, and other promoters lack a TATA-box but are INR-dependent (Martinez et al. 1994, 1995).

TATA-box type promoters are less common in vertebrates (Cooper et al. 2006; Carninci et al. 2006), and more than three-quarters of human core promoters lack a TATA box or TATA-like sequence (Yang et al. 2007). Of these TATA-independent promoters, 30 % are estimated to contain an INR sequence, but ~46 % of human promoters lack both motifs. In addition, some promoters containing INR motifs are of the dispersed type, so the percentage of dispersed promoters is likely to be higher than this estimation. Evolution has clearly favoured dispersed promoters that produce greater variation of expression due to a larger number of interacting factors and possibilities. Dispersed promoters are also more malleable than focused promoters, as new elements can be incorporated easily without disturbing the general functions of the promoter.

The low prevalence of TATA-box containing genes in higher eukaryotes indicates that CpG promoters have been evolutionarily selected for to facilitate large-scale complex gene expression. However, what about the TATA-box promoters that remain? Promoters containing a TATA box and INR are over-represented in genes involved in nucleosome assembly and cell adhesion, while those containing a TATA-box only are over-represented in genes involved in cellular responses and organogenesis (Yang et al. 2007). This selectivity suggests that the TATA-box containing promoters are often associated with cell-type specific genes, and in correlation with this, housekeeping genes usually contain the more complex CpG promoters (Yang et al. 2007). This can be explained by the need for housekeeping genes to be ubiquitously expressed in all cell types. Different cell types have varying populations of transcription factors, microRNAs, protein factors and other regulatory elements that interact with DNA and mRNA to alter gene expression. If a housekeeping gene is to be constitutively expressed, it is important that variations in regulatory elements do not result in insufficient or over-expression of these genes. Therefore it makes sense that housekeeping gene promoters would be of the dispersed kind, where multiple factors work together to maintain the correct balance. For cell-type specific genes, where the factor population is expected to remain largely the same, a focused promoter could ensure swift and consistent gene expression within the specific cell type. Interestingly, TATA-box containing promoters generally have significantly lower GC content (45–50 % GC) than those without a TATA-box (60 % GC), and the base composition is similar to the AT-rich promoters found in Drosophila. AT-rich sequences are expected to unwind more easily due to stronger base-stacking interactions in GC-rich sequences, but it is not known whether this is the reason for the higher AT concentration seen in TATA-box containing promoters. The sequence composition of focused promoters and the similarity with drosophila promoters provides further evidence that TATA-boxes are ancient conserved motifs and that focused and dispersed promoters fulfil different requirements of the regulatory network.

2.1.2 Dispersed Promoters

Although the mechanism of TATA-box and INR initiated transcription are well characterised, the more common type of promoter in humans are dispersed promoters that are more complex and consequently less well characterised. Dispersed promoters are generally GC-rich, and often contain multiple Sp1 transcription factor binding sites (GC boxes with the consensus GGGCGG, aka M6) (Smale and Kadonaga 2003). The presence of multiple binding sites is important and is a good example of the way dispersed promoters function in a complex manner. Multiple Sp1 proteins can bind in various places within the promoter at the same time. Three different isoforms of Sp1 have been identified, and post-translational modifications such as phosphorylation can have a significant impact on Sp1 activity, which can act to enhance or repress transcription. Other potential motifs involved in transcription initiation identified are M3 (ELK-1) and M22, although little is known about these sites and the elements that bind to them (Yang et al. 2007).

Due to the characteristics of Sp1 and its importance in transcription initiation, it is not surprising that GC-rich promoters are the more prevalent type in humans. *In silico* analysis of the average GC content of 15,685 human promoter regions in the UCSC GoldenPath database found that the region −250 to +250 ranged from 55 to 60 % GC (Yang et al. 2007). A significant contribution to the GC content is from CpG islands. CpG sites occur when a G follows a C on the same strand of DNA or RNA, joined by a phosphodiester bond. CpG islands consist of multiple CpG sites and range in size from 200 to 3,000 nucleotides (reviewed in Deaton and Bird 2011). CpG sites not contained within CpG islands are normally sites for DNA methylation. Catalysed by DNA methyltransferase, methylation generally represses transcriptional activity. However, it seems that methylation does not usually occur at CpG sites located within islands.

While CpG islands are a well-characterised feature of dispersed promoters, about 50 % of the CpG islands in the genome are not associated with an annotated promoter (termed "orphan" CpG islands) (Illingworth et al. 2010). However, there is evidence of transcription initiation at many orphan islands, so they are likely to represent either uncharacterised promoters of genes or promoters driving transcription of noncoding RNA. Importantly, methylation often does occur at these orphan CpG islands, causing repression of transcription. This could mean that these are real promoters but encode genes in a cell-type or developmentally specific manner, such that in many contexts the gene is not expressed (Illingworth et al. 2010). It appears the scope of promoters is less well characterised than previously thought, and more research is required to elucidate the function of these orphan CpG islands.

2.2 Complex Factor Interactions

Genes with complex promoters are likely to selectively make use of regulatory elements, such as enhancers and silencers, allowing varying levels of expression as required. The *IFN-beta* enhancer element has been demonstrated to "loop out" the intervening DNA to access the promoter (Nolis et al. 2009). This allows specific control of gene activation (i.e. via a gene specific enhancer) using general factors. The conformation of the TFIID complex also appears to differ when it is bound to different core promoters, allowing interaction with a large range of subsets of transcriptional activators (Smale and Kadonaga 2003). A recent study of non-prototypical core promoter recognition factors identified a number of cell-type-specific factors that act in potentiating developmental gene regulation and cellular differentiation (Goodrich and Tjian 2010). In addition, promoter-selective homologues of basal transcription factors and considerable diversity in the sequence structure and composition of core promoter elements allows complex programs of tissue-specific and promoter-selective transcription, potentially producing a number of specifically expressed gene isoforms (Davuluri et al. 2008). These studies show that promoters in higher organisms are complex regulatory regions consisting of multiple binding elements that can recruit a variety of *cis*-acting regulatory factors as required by the cell. This also has implications for interactions between unrelated genes that are regulated by the same factors, as factor binding to one gene could restrict the availability to other genes. It is clear that eukaryotic gene expression exists in a fluid system in which balance between factors is important.

2.3 Alternative Promoters

Promoter usage can have a major impact on gene expression and many mammalian genes contain multiple promoters (Cooper et al. 2006). Alternative promoter use is a widespread phenomenon in humans (Cooper et al. 2006) that can alter expression of the associated gene at both the mRNA and protein level. It is also an important mechanism involved in the cell-specific or developmental-specific expression of many genes (Levine and Tjian 2003). For example, TATA-box-deficient and TATA-box-containing alternative promoters of the hemoglobin γ A gene (*HBG1*) are used during and after embryonic development, respectively (Duan et al. 2002), showing that the basal transcription apparatus can be recruited to different types of core promoters in a developmental stage-specific manner (Davuluri et al. 2008). Another more recent example demonstrates the complexity and variation that can arise through the use of alternative promoters for regulation of the *MITF* transcription factor during vertebrate eye development. Each of the nine alternative promoters associated with expression of this gene produce isoforms containing different first exons and protein binding sites, allowing variable spatial and temporal expression of different protein isoforms during the complex process of eye development (Bharti et al. 2008). A recent global analysis of mammalian promoters concluded that alternative promoters are over-represented among genes involved in transcriptional regulation and development, which makes sense because alternative promoters are likely to be utilised to alter the expression of a gene in different contexts, while single-promoter genes are active in a broad range of tissues and are more likely to be involved in general cellular processes, such as RNA processing, DNA repair, and protein biosynthesis (Baek et al. 2007).

Alternative promoter usage has also been implicated in the production of biologically distinct protein isoforms (Davuluri et al. 2008). Lymphoid enhancer factor (*LEF1*) is transcribed from two alternative promoters; promoter 1 produces a full length isoform that activates target genes *Wnt/β-catenin*, while promoter 2, situated in the first intron, produces a shorter isoform that represses target genes (Arce et al. 2006). The use of alternative promoters will also affect the 5'UTR, which can alter the stability or translation efficiency of the mRNA variants while encoding identical proteins. Short stature homeobox (*SHOX*), a cell-type specific transcription factor involved in cell cycle and growth regulation uses two alternative promoters producing two distinct 5'UTRs (one is longer and highly structured), resulting in identical proteins that are regulated differently by a combination of transcriptional and translational control mechanisms (Blaschke et al. 2003). The regulatory effect of the 5'UTR will be discussed in more detail in the next section. These examples confirm that alternative promoter usage can play a major role in the spatial and temporal control of gene expression and that use of alternative promoters is an effective way of increasing the complexity of gene expression pathways.

2.4 Bidirectional Promoters

How promoter selection is determined is not fully understood, but possible mechanisms of promoter switching include diverse core-promoter structure at alternative promoters, variable concentration of *cis*-regulatory elements in the upstream promoter region and regional epigenetic modifications, such as DNA methylation, histone modifications and chromatin remodelling (Davuluri et al. 2008). In addition to multiple promoters and promoter-like elements, it is now clear that bidirectionality is a common feature of promoters, with extensive analyses performed in yeast (Lin et al. 2007; Xu et al. 2009) and human (Lin et al. 2007), with an estimated ~11 % of human genes expressed via bidirectional promoters. To date, the impact of this is not known, but it is suggested that bidirectional transcription has a role in maintaining an open chromatin structure at promoters, and may also provide a mechanism to spread the transcriptional regulatory signals locally in the genome or play a role in the coordinated expression of gene networks (Xu et al. 2009). This relatively new finding has implications for the complexity of the transcriptional network, and future research is likely to uncover more evidence of bidirectionality or interacting promoter regions.

Co-regulation is a consequence of bidirectional promoters, and how these promoters act to ensure both genes are efficiently and stably transcribed is only now being discovered. In mammals, two genes encoding mitoribosomal protein S12 (Mrps12) and the mitochondrially localized isoform of seryl-tRNA ligase (Sarsm or Sars2) share a conserved bidirectional promoter region of <500 bp (Yokogawa et al. 2000; Shah et al. 2001). A study by Zanotto et al. (2007) identified four adjacent CCAAT box elements within this promoter region that interact with the transcription factor NF-Y, which is capable of recognizing the core binding sequence in either orientation, and will thus also recognize ATTGG on the opposite strand (Mantovani 1998). ChIP analysis, *in vivo* footprinting, electrophoretic mobility shift assay and reporter analysis revealed that NF-Y binding at all four CCAAT sites confers varying transcriptional selectivity (Zanotto et al. 2007). CCAAT box 1 (closest to Mrps12) and box 4 (closest to Sarsm) show a good match for NF-Y consensus binding and each confers transcriptional bias for the adjacent gene. CCAAT boxes 2 and 3 have weaker binding capacities and are likely to function as accessory elements (Zanotto et al. 2007). Reporter assays using contructs lacking the coding region and 3'UTR indicated a transcriptional bias in the Mrps12 direction resulting in fourfold higher expression. However, steady state mRNA levels in cultured 3T3 cells were similar for Sarsm and Mrps12. This is indicative of post-transcriptional regulation, and demonstrates the limitations of reporter assays in studying gene expression. Although reporter assays are useful for studying promoter activity, there are numerous other factors regulating gene expression, and this assay only gives an insight into part of the process. Nevertheless, this study has confirmed the bidirectional capabilities of NF-Y, and it is likely that varying levels of binding to each CCAAT site allows directional bias according to the requirements of the cell. Other promoter elements were identified, NRF-2 and AP-1, that enhance transcription but do not confer directional bias.

Bidirectional promoters may also be employed to allow coregulation of genes that have similar functions. The sirtuin 2 gene family (*SIR2*) are important genes involved in metabolism and aging. In humans, a homologue of *SIR2, SIRT3*, sits adjacent, but in the opposite orientation, to the proteasome 26S subunit non-ATPase 13 gene (*PSMD13*) that encodes the p40.5 regulator subunit of the 26S proteasome (Bellizzi et al. 2007). The proteasome functions to degrade abnormal proteins, and thus also plays a role in aging. Investigation into the promoters revealed that both genes are regulated by a bi-directional promoter, consisting of common core Sp1 sites (Bellizzi et al. 2007). Linkage disequilibrium studies revealed that variability in the expression of PSMD13-SIRT3 was different in very old people compared to younger people. Interestingly, the distance between the two genes, and thus the length of the promoter region, was found to have increased throughout evolution, (e.g. from 86 bp in mouse to 788 bp in humans), which is indicative of an increase in the complexity of the human promoter (Bellizzi et al. 2007).

2.5 Conclusion

It is evident that eukaryotic promoters have evolved from the relatively simple 'switches' found in bacteria, to the complex multi-factor regulatory regions found in mammals today. Complex promoters induce a range of responses to varying environmental conditions and cellular signals, facilitating controlled expression of the required gene variant according to developmental stage and cell type. Control of this kind is the basic requirement for producing the complex expression patterns necessary for cellular differentiation, and thus for the development of complex organisms.

3 5′ Untranslated Region

The 5′ untranslated region (Anastasi et al. 2008) is a regulatory region of DNA situated at the 5′ end of all protein-coding genes that is transcribed into mRNA but not translated into protein. 5′UTRs contain various regulatory elements (Fig. 1b) and play a major role in the control of translation initiation. Here, we discuss the regulatory roles of the 5′UTR, highlighting how the number and nature of regulatory elements present, as well as the secondary structure of the mRNA and factor accessibility, impact upon the expression of the downstream open reading frame (Bradnam and Korf 2008).

3.1 Structure

3.1.1 5′ Cap Structure

The 5′ cap is a modification added to the 5′ end of precursor mRNA that consists of 7-methylguanosine attached through a 5′-5′-triphosphate linkage, reviewed in

(Banerjee 1980). This structure is essential for efficient translation of the mRNA, serving as a binding site for various eukaryotic initiation factors (eIFs) and promoting binding of 40S ribosomal subunits and other proteins that together make up the 43S pre-initiation complex (PIC) (Jackson et al. 2010). In addition to promoting translation, a recent study showed that the triphosphate linkage of the 5′ cap inhibits mRNA recruitment to the PIC in the absence of the full set of eIF factors (Mitchell et al. 2010). The authors suggest that this mechanism allows inhibition of non-productive recruitment pathways, preventing the assembly of aberrant PICs that lack the factors required for efficient scanning and translation initiation (Mitchell et al. 2010). The 5′ cap structure also functions in stabilisation of the mRNA, with the initiation of mRNA decay reliant on the removal of the cap by various de-capping enzymes (Meyer et al. 2004). Although the major role of the 5′ cap seems to be the facilitation of mRNA translation, recent investigations of non-coding RNAs revealed that some types of non-coding RNAs, such as promoter-associated-RNAs (PASRs) are also capped (Fejes-Toth et al. 2009). The role of the cap in the regulation of these transcripts is currently unknown, and further studies are likely to reveal additional regulatory roles for this structure.

3.1.2 Secondary Structure

The structure and nucleotide content of the 5′UTR appears to play an important role in regulating gene expression, with genome-wide studies revealing marked differences in structure and nucleotide content between housekeeping and developmental genes (Ganapathi et al. 2005). In general, 5′UTRs that enable efficient translation are short, have a low GC content, are relatively unstructured and do not contain upstream AUG codons (uAUGs), as revealed by in silico comparisons of genes with low and high levels of protein output (Kochetov et al. 1998). In comparison, 5′UTRs of genes with low protein output are, on average, longer, more GC rich, and possess a higher degree of predicted secondary structure (Pickering and Willis 2005). These highly structured 5′UTRs are often associated with genes involved in developmental processes, and the corresponding mRNAs are usually expressed in a developmental or tissue-specific manner. This variation in expression is likely to be mediated by interactions with different RNA binding proteins and structural motifs within the 5′UTR region. For example, the peroxisome proliferator-activated receptor γ (PPAR-γ) gene expresses a number of splice variants that differ in the 5′UTR rather than the protein-coding domain. Analysis of the translational activity of the various 5′UTRs found three that enhanced translation and two that had a repressive effect (McClelland et al. 2009). MFOLD modelling of mRNA folding in the 5′UTR revealed the presence of compact structures around the start codon in the repressive 5′UTRs. Although the exact mechanism of repression is unknown, it is likely that the differences in the structure and nucleotide content of the 5′UTRs facilitate binding of different proteins that act to either enhance or repress translation.

Although there are some general trends regarding 5′UTR, it is important to remember that assumptions cannot be made purely by looking at the sequence characteristics. While longer 5′UTRs are often associated with genes of low or selective expression, the huge number of parameters influencing transcription and translation requires experimental validation before any conclusions can be made.

3.1.3 G-Quadruplexes

A well-characterised secondary structure that has a major impact on translation is the G-quadruplex structure (G4). These structures are guanine-rich nucleic acid sequences that can fold into a non-canonical tetrahelical structure that is very stable and has the ability to strongly repress translation (Beaudoin and Perreault 2010). Bioinformatic studies have shown that these structures are often highly conserved, can be found in regulatory elements other than the 5′UTR, such as promoters, telomeres and 3′UTRs, and are enriched in mRNAs encoding proteins involved in translational regulation and developmental processes, indicating that they are an integral part of various important biological processes (Beaudoin and Perreault 2010). Many G4 structures have also been found in oncogenes. The *TRF2* gene, which is involved in control of telomere function, has a G-rich sequence within its 5′UTR that can fold into a G4 structure and repress translation of a reporter gene by 2.8-fold (Gomez et al. 2010). This gene is overexpressed in a number of cancers, indicating that the G4 is in place to tightly regulate the expression of this gene. Gomez and colleagues also demonstrated that a number of ligands that bind to G4 structures were able to modulate the translation efficiency of *TRF2 in vitro* (Gomez et al. 2010). In conclusion, G4 s appear to have a major impact on the translational regulation of the genes in which they reside (Beaudoin and Perreault 2010) and may repress translation by secondary structure alone or by modulating interactions with proteins and other factors.

The expression of the *NRAS* proto-oncogene is another example of a gene controlled by a G4 element, which is an 18nt element situated close to the 5′cap within the 5′UTR (Kumari et al. 2007). A reporter plasmid was used to investigate the effect of the G4 and its position in the 5′UTR on translation (Kumari et al. 2008). The study found that when the G4 was situated close to the 5′UTR (positions +2, +14 and +47), the translation efficiency was reduced by more than 50 %. However, insertion of the G4 at positions +120 and +233 seemed to have no effect on translation (Kumari et al. 2008). The *NRAS* G4 contains three G-tetrad structures. Constructs were designed containing either 2 or 4 tetrads to compare the effect on translation. *In vitro* experiments demonstrated that reducing the number of tetrads to 2 increased the translation efficiency twofold. The construct containing 4 tetrads did not have a significant impact on translation, but UV-melting analysis indicated this structure was not significantly more stable than the 3-tetrad G4. This data shows that the stability of the G4 can have a modulating effect on translation (Kumari et al. 2008). The position of this G4 in the *KRAS*

5′UTR is a good example of a G4 acting to tightly control gene expression of an important regulatory gene.

The scanning model of translation initiation proposes that upon binding to the 5′ cap, the 43S ribosome complex scans the 5′UTR until it locates the optimal AUG codon and initiates translation (Kozak 1989). This model led to an assumption that all mRNAs with highly structured 5′UTRs have low translation rates due to inability of the ribosome to scan through tight secondary structures such as stem-loops. However, some recent studies have shown that this is not the case. Firstly, a report (Dmitriev et al. 2009) highlighted the limitations of the previously preferred analysis method used by many groups, the rabbit reticulocyte lysate (RRL) system (Pelham and Jackson 1976). In a comparison of methods for studying translation, they found the RRL system possessed a number of flaws, the most important of which was that capping did not seem to significantly affect translation when using this cell-free system. As it is well established that the 5′ cap is essential for efficient translation and that the effect of the 5′ cap is much more pronounced for some mRNAs compared to others, the RRL system seems not to reflect *in vivo* conditions (Shatsky et al. 2010). In addition, correlating evidence from experiments using a different cell-free system (wheat germ S30 system) and cultured cells demonstrated that capping increased the translational efficiency for most RNAs by several orders of magnitude (Dmitriev et al. 2009). Importantly, using these two systems, Dmitriev and colleagues found no dramatic difference in the translational efficiency between several short, unstructured and longer, highly structured 5′UTRs that they examined in their study. This data indicates that the natural stem-loop structures in these 5′UTRs do not seem to inhibit initiation. Despite this, large-scale in silico studies have shown there is a significant correlation between 5′UTR folding free energy and protein abundance (Ringner and Krogh 2005). This does not mean that the structure itself is the inhibitory factor, although it does suggest that 5′UTR secondary structure is involved in post-transcriptional regulation.

It has been emphasised that interactions with RNA-binding proteins prior to scanning and initiation are likely to affect the mechanism of searching for the initiator codon (Dmitriev et al. 2009). For example, the eIF4F complex assembles on the 5′ cap prior to translation and unwinds secondary structures in the 5′UTR in order to promote loading of the 43S ribosomal complex onto the mRNA (Kapp and Lorsch 2004). This correlates with the results obtained by Dmitriev and also helps explain why direct inhibition via secondary structures is observed in the RRL system, as this system has a highly reduced content of mRNA-binding proteins (Svitkin et al. 1996). The human L1 bi-cistronic mRNA contains a 900-nt long 5′UTR with high GC content (~ 60 %) and two short upstream open reading frames (uORFs). Predicted folding reveals a number of potential stem-loop structures, however the L1 mRNA is still translated very efficiently via cap-dependent initiation (Dmitriev et al. 2007). The above examples provide strong evidence that the unwinding of stem-loops occurs sequentially and indicates that the current practice of using in silico predictions of folding energies of 5′UTRs to forecast translatability is likely to result in incorrect assumptions. Stem-loop structures may have the ability to inhibit translation but this is more likely to occur via protein binding, rather than inhibition by the mRNA structure itself, except in the case of very stable structures such as G4 s. The collagen α1(I) mRNA contains a 5′ stem-loop structure that inhibits translation *in vitro* (Stefanovic and Brenner 2003). This study demonstrated that when the stem-loop structure is mutated *in vivo*, the collagen that is translated is pepsin sensitive, compared to wild type translation that results in the production of pepsin resistant collagen. In addition, the wild-type transcript produces disulfide bonded high molecular weight collagen, production of which is almost abolished following mutation of the stem-loop. These data indicate that the 5′ stem-loop in collagen α1(I) is required for stabilisation of the collagen triple helix structure, a process most likely mediated by stem-loop interacting proteins (Stefanovic and Brenner 2003).

3.1.4 Alternative 5′UTRs

In addition to those UTRs generated via the use of alternative promoters, alternative 5′UTRs may be produced by alternative splicing or through usage of alternative transcription start sites from a single promoter (Smith 2008). Diversity within the 5′UTR of a gene enables variation in expression, depending upon the nature of the regulatory elements contained within each alternative 5′UTR. Slight changes in the arrangement of translational control elements between isoforms can lead to major changes in the regulatory effects on translation (Resch et al. 2009). A large-scale analysis of the mammalian transcriptome indicates that expression of alternative 5′UTRs is a widespread phenomenon, with most genes having the potential for differential expression (Hughes 2006). Genes that are known to consistently express multiple 5′UTRs are typically involved in core functional activities such as transcription and signalling pathways (Resch et al. 2009). The oestrogen receptor β gene (*ERβ*) plays an important role in oestrogen function and the expression of the multiple isoforms is frequently mis-regulated in cancers. Smith and colleagues have recently identified three alternative 5′UTRs (termed UTR a, c and E1) that contribute to the expression of the different isoforms (Smith et al. 2010a, b). They found that UTRs a and c inhibited translation, with UTRa having a very potent inhibitory effect, while E1 had a less pronounced, but still inhibitory effect, despite being only 90nt long and having low predicted secondary structure.

The growth hormone (GH) receptor is produced by a gene that also generates a GH-binding protein by proteolytic cleavage of the GHR (Sotiropoulos et al. 1993). The gene is encoded by exons 2–10. Exon 1 has nine alternative exons, coding for alternative 5′UTRs (Goodyer et al. 2001b) that regulate the expression of GHR, facilitating differential expression among tissues and throughout development, to modulate the response to GH (Southard et al. 1995). Interestingly, two primate-specific first exons were identified that contain Alu-elements (Goodyer et al. 2001a). The alternative first exons of the GHR gene have been mapped to two main clusters, (38 kb and 18 kb upstream of exon 2), although one has been identified between the two clusters and another close to exon 2. This means that the GHR gene is also under the control of multiple promoters. This highly complex transcriptional unit and its evolution from rodents to primates is a fine example illustrating the nature of regulatory systems often required for expression of essential genes in higher eukaryotes. The expression of alternative 5′UTRs represents an evolutionary gain of transcriptional and translational control pathways, allowing tissue-specific expression patterns and expanding the repertoire of expression from a single gene locus.

3.2 Regulatory Motifs

The lack of correlation between the rate of translation and the length or structure of the 5′UTR in both capped and uncapped mRNAs, as well as the ability of certain genes to be expressed under conditions of stress indicates that there must be other elements within eukaryotic mRNAs that contribute to translation initiation and control of gene expression via the 5′UTR.

3.2.1 IRES and Cap-Independent Translation Initiation

Internal ribosome entry sites (Birney et al. 2007) are mRNA regulatory motifs that facilitate a cap-independent mechanism of translation initiation, in which the ribosome binds to an internal site close to the translation initiation site (Meijer and Thomas 2002). IRES allow recruitment of ribosomes to capped or uncapped mRNAs under conditions when cap-dependent translation is inhibited by stress, cell-cycle stage or apoptosis, ensuring the continued expression of essential proteins required for cell function. A number of IRES-containing genes such as *c-Myc*, *Apaf-1*, and *Bcl-2* are required at low levels during normal cellular growth, but are induced via the IRES pathway under conditions of stress (Komar and Hatzoglou 2005). It is thought the IRES pathway may also contribute to maintaining the low expression levels required under normal cellular conditions by sequestering ribosomes and reducing their binding at the main translation initiation site. The mechanism of internal initiation is still poorly understood, although it is clear that efficiency of IRES is heavily reliant upon *trans*-acting protein factors, allowing cell-specific IRES-mediated translation of mRNAs (Pickering and Willis 2005).

Structures in the 5′UTR have been shown to influence IRES activity, which may occur via interactions with various trans-acting factors, or by direct interactions with ribosomes. An example of genes in which IRES activity is regulated by *trans*-acting factors is the Myc family of proto-oncogenes that are involved in cell proliferation. Recruitment of ribosomes to the IRES is dependent upon at least four proteins that bind and alter the conformation of the mRNAs to allow interaction with the 40S subunit (Cobbold et al. 2008). Another example is the Hepatitis C virus (*HCV*), containing a highly structured IRES that initiates cap-independent translation via two major structural domains, consisting of conserved stem-loop structures that interact with the 40S ribosomal subunit to form a complex and recruit eIF3 (Lukavsky 2009). The structures of eukaryotic IRES are very diverse and no universally conserved sequences or structural motifs have yet been identified. For some genes, specific and stable RNA structures are required for efficient IRES activity, while in other genes, stable structure is inhibitory to IRES-mediated translation (Filbin and Kieft 2009). It has been suggested that IRES are not rigid structures but can undergo transitions that substantially influence their activity (Komar and Hatzoglou 2005). IRES elements may also result in the production of different protein isoforms, thus further expanding the repertoire of expression from a single gene (Komar and Hatzoglou 2005).

The presence of IRES between different AUG and non-AUG initiation codons suggests a role for IRES in promoting translation initiation from weak alternative start codons (Touriol et al. 2003). IRES may also interact with uORFs, another class of regulatory elements discussed in the next section. Gilbert (2010) discusses recent findings on IRES and draws attention to flaws in the methods for defining IRES (bicistronic test) that may result in false positive predictions (Gilbert 2010). Although IRES are an important mechanism for some genes, Gilbert suggests that it is wrong to assume the presence or activity of an IRES by prediction alone, emphasizing the importance of experimental validation. IRES are a poorly understood but important regulatory mechanism, and further investigation will be needed to discern the mechanisms and context of initiation via IRES.

The X chromosome-linked inhibitor of apoptosis (*XIAP*) is an important regulator of apoptosis and therefore its expression must be tightly controlled. The *XIAP* 5′UTR is 1.7 kb in length and contains an IRES that facilitates synthesis of XIAP protein under conditions of cellular stress (Holcik et al. 1999). In conjunction with IRES-mediated regulation, a recent study revealed the presence of an *XIAP* mRNA species with a smaller 5′UTR generated by alternative splicing (Riley et al. 2010). This 323nt 5′UTR does not contain the IRES and was found to be present at a much higher level (10×) than the longer 5′UTR. Vector studies indicate that the majority of expression from the longer 5′UTR is mediated by the IRES. In addition, translational studies demonstrated that the shorter 5′UTR has a much lower translation efficiency compared to the longer 5′UTR (Riley et al. 2010). The shorter 3′UTR is responsible for high expression of XIAP under normal growth conditions, in a cap-dependent manner. On the other hand, the longer 3′UTR contains an IRES that allows cap-independent translation under conditions of stress, as demonstrated by serum starvation (Riley et al. 2010). The combination of alternative 5′UTRs and IRES-mediated translation ensures protein expression at all times.

Another example of control of gene expression via a complex 5′UTR region is the human fibroblast growth factor 2 gene (*FGF-2*). *FGF-2* is expressed as five different isoforms resulting from alternative initiation codons within the 5′UTR, and translation can also occur from an IRES (Bonnal et al. 2003). Interestingly, translation from four of the five initiation codons is mediated by the IRES. The authors suggest that the IRES facilitates translation at each of the four codons via modulation of the RNA structure by trans-acting factors (Bonnal et al. 2003).

3.2.2 uORFs

Upstream open reading frames occur in 5′UTRs when an in-frame stop codon follows an upstream AUG (uAUG) codon, prior to the main start codon (reviewed in Mignone et al. 2002; Morris and Geballe 2000; Wethmar et al. 2010). uORFs are present in approximately 50 % of human 5′UTRs and their presence correlates with reduced protein expression and with mutation studies indicating that on average, uORFs reduce mRNA levels by 30 % and reduce protein expression by 30–80 % (Calvo et al. 2009). Ribosomes binding to the uAUG may translate an uORF, which can impact on downstream expression by altering the efficiency of translation or initiation at the main ORF. If efficient ribosome binding does not occur, the result will be a reduction of protein expression from the gene. Alternatively, synthesis may continue from the uORF and produce an extended protein that may be detrimental. Decreased translational efficiency is a well characterised effect of uORFs within a 5′UTR (Morris and Geballe 2000), illustrated by the Poly(A)polymerase-α (*PAPOLA*) gene that contains two highly conserved uORFs in the 5′UTR. Mutation of the 5′ proximal uAUG codon resulted in increased translation efficiency, indicating that the uORF has a significant inhibitory effect on the expression of this gene (Rapti et al. 2010). Another example is the thyroid hormone receptor that represses and activates transcription of a number of target genes, in the absence or presence of thryoid hormone, and is strongly repressed by a 15nt uORF in its own 5′UTR (Okada et al. 2012). It is commonly thought that uORFs decrease translational efficiency by rendering the ribosome unable to reinitiate translation following termination from the uORF (Meijer and Thomas 2002). However, a recent study of over 500 uORF-containing gene loci found no significant correlation between the impact of the uORF on the expression of the downstream gene and the distance between the uORF and the coding sequence (CDS) (Calvo et al. 2009). The authors suggest it is likely that in genes containing a single uORF, CDS translation occurs from ribosomes that scan through the uORF, rather than via re-initiation. This is in contrast to the work of Kozak (1987), and the general consensus on uORFs. To further complicate matters, experiments using cells depleted of Rent1, a factor involved in nonsense mediated decay (NMD), revealed that in the absence of NMD, transcripts containing uORFs were generally upregulated (Mendell et al. 2004). This implies that NMD also plays an important role in the regulation of these transcripts. It seems that uORFs present a number of options for the ribosome, and whether translation occurs, scanning continues or reinitiation at the main ORF occurs, depends on a number of factors. The yeast *GCN4* transcriptional activator contains four uORFs in the 5′UTR, only one of which allows reinitiation of the ribosome at the main ORF. *Cis*-enhancers on either side of the uORF allow efficient reinitiation. Investigation showed that the 5′ *cis*-acting sequences interact with the N-terminal domain of the eIF3a/TIF32 subunit of the initiation factor eIF3 to stabilize post-termination 40S subunits on uORF1 so the ribosome remains attached to the transcript and continues scanning downstream (Munzarova et al. 2011). A combination of in silico and experimental data suggest that reinitiation is facilitated by specific mRNA folding during ribosome scanning that allows a sequence of interactions to occur (Munzarova et al. 2011). This is just one example but it is clear that the mechanism of uORF gene knockdown is more complex than the scanning model proposes. Further experimentation will be required to elucidate this mechanism or mechanisms.

AUG codon recognition is influenced by a number of factors, including proximity of the AUG to the 5′ cap, the flanking sequence and secondary structure (Krummheuer et al. 2007). uORFs appear to exist as regulatory elements that act to control the translation of the downstream ORF. Protein kinase C (*PKC*) represents a family of serine/threonine kinases that play a major role in the regulation of cell growth and differentiation (Raveh-Amit et al. 2009). The novel PKCη isoform has a specific tissue distribution and is primarily expressed in cells undergoing high turnover, such as epithelial cells. Recent studies found that this isoform has a special role in the response to stress and its expression has been found to correlate with drug resistance in various cancer types (Rotem-Dai et al. 2009). The 5′UTR of human *PKCη* is long (659nt), GC rich, and contains two small conserved uORFs (Raveh-Amit et al. 2009). Mutations introduced into each of the uORFs resulted in modest increases in expression (1.5- and 2.2-fold increases) and a double mutation resulted in a threefold increase in gene expression from the main AUG. This mechanism of translational repression is likely to be in place to control the expression of *PKCη* under normal cellular conditions (Raveh-Amit et al. 2009). Under conditions of stress (e.g. Glucose deprivation or hypoxia), the two uORFs also play a role in expression, as they facilitate leaky scanning to enhance the translation of the main ORF. Varying levels of ribosome binding and translation of each of the uORFs may also contribute to cell-specific 'tweaking' of gene expression.

In addition to the AUG consensus, studies have shown that all codons differing from AUG by a single nucleotide can be utilised for translation initiation (Peabody 1989), the efficiency of which depends on the sequence context. Orthinine decarboxylase (ODC) is involved in the biosynthesis of polyamines and its expression is modulated through an uORF with an AUU codon (Ivanov et al. 2008). Initiation at the AUU uORF varies, depending on the cellular concentration of polyamines, with initiation at the AUU significantly reduced in polyamine-depleted cells, where initiation is only 18 % as efficient as at the main ORF, compared to 54 % in polyamine-supplemented cells. This is consistent with reduced expression of ODC when polyamines are present, and is a very logical way of controlling gene expression: modulating expression relative to the concentration of interacting or environmental factors. This example demonstrates that there are likely to be greater numbers of uORFs than previously predicted.

Despite the majority of uORFs having a negative impact on gene expression, there are some cases in which the presence of a uORF actually enhances translation. Bicistronic *vpu-env* mRNAs are involved in HIV-1 virus expression and they contain a conserved minimal uORF (Krummheuer et al. 2007). This uORF is only 5nt upstream of the *vpu* AUG and is immediately followed by a termination codon that overlaps the main AUG. Krummheuer and colleagues showed that this uORF has a significant positive impact on the translation of *Env*, while not interfering with translation of *Vpu* (Krummheuer et al. 2007). Mutants in which the distance between the uORF and the main AUG was increased by 5 codons indicated that the uORF is not involved in the initiation of *Vpu*, and the authors suggested that the minimal uORF may act as a site for ribosome pausing, allowing it to interact with an RNA structure that supports a ribosome shunt, a process during which the ribosome physically bypasses part of the 5′UTR to reach the initiation codon.

The role of uORFs as regulatory elements acting on the process of ribosome binding and translation is well studied, but the function or fate of the encoded peptides is often unknown, perhaps due to the difficulty in analysing the expression levels and localisation of the peptides. Evidence that peptides translated from uORFs are present in cells was first shown by Oyama and colleagues (2004), who identified 54 proteins of <100 amino acids expressed in human chronic myelogenous leukemic cells that were all mapped back to uORFs (Oyama et al. 2004). Although proteins were identified, thousands of uORFs did not seem to produce a detectable protein product in these cells, which indicates that either (i) proteins derived from uORFs may be selectively proteolyzed in the cells, (ii) some of the uORFs are expressed but not in this cell type, or (iii) many do not produce proteins. Despite this, it is clear that some uORFs do produce peptides that are retained in the cell and thus are likely to be functional, although to date there are no comprehensive studies on the function of proteins translated from an uORF.

The past decade has revealed that regulation via uORFs is a complex process that acts to tightly regulate the expression of the genes they control. A good example of complex control of gene expression via uORFs was outlined recently by Suzuki et al. (2010). RNase H1 is present in the nuclei and mitochondria of mammalian cells and is differentially expressed among cell types. Two different in-frame AUGs control the expression of these isoforms and an uORF is also present in the 5′UTR of this gene. Translation of the mitochondrial RNAse H1 was found to be initiated at the first AUG, which is restricted by an uORF, resulting in the mitochondrial isoform being about 10 % of the abundance of the nuclear form (Suzuki et al. 2010). Translation of the nuclear isoform proceeds from the second AUG and is unaffected by the presence of the uORF, as the ribosome either efficiently reinitiates or skips both the first AUG and the uORF. This regulation allows control of RNase H1 expression in mitochondria, where its excess or absence can lead to cell death, without affecting the normal expression levels of the nuclear isoform. Suzuki and colleagues also found that altering the context of the AUG altered transcript accumulation, meaning there must be other factors involved. This example illustrates the combinatorial use of multiple uORFs and other factors to produce a highly specific system of translational regulation. In addition, alternative promoters or splicing, as well as the finding that out-of-frame and sub-optimal initiation codons can, in certain contexts be available to ribosomes, and are all factors that can affect uORF expression, further increasing the diversity of regulation and translation emerging from these regions (Oyama et al. 2007). A recent study that treated a human monocytic cell line with puromycin, to prematurely stop translation, before utilising ribosomal footprinting to identify translation initiation sites indicated that uORFs are more prevalent than predicted by in silico data (Fritsch et al. 2012). Their experiments predicted 2,994 novel uORFs in 5′UTRs alone, with more predicted overlapping the coding regions and 3′UTRs. This emphasizes the importance of using *in vitro* and *in vivo* techniques to investigate the presence of active uORFs.

Mutations involving uORFs are likely to be detrimental, as they can disrupt the control of gene expression, resulting in aberrant gene expression levels that may subsequently lead to disease (Chatterjee and Pal 2009). Mutations disrupting the uORF in the 5′UTR of the gene encoding the human hairless homolog (*HR*) and resulting in increased translation of the gene, have been associated with Marie Unna hereditary hypotrichosis, an autosomal dominant form of genetic hair loss (Wen et al. 2009). Mutations that create novel uORFs may also have a detrimental effect by interfering with normal expression. It has been speculated that a mutation in a tumour suppressor gene may result in decreased production of protective proteins and contribute to the onset of cancer (Wethmar et al. 2010). Mutations in *CDKN2A* that encodes a kinase inhibitor have been associated with predisposition to inherited melanoma. A G > T in the 5′UTR gives rise to an uORF that decreases translation from the main initiation codon was associated with inherited melanoma cases (Liu et al. 1999). These examples illustrate the importance of uORFs in the control of specific gene expression and in maintaining homeostasis, and variability within uORFs is thought to contribute to individual phenotype and disease susceptibility (Wethmar et al. 2010).

3.3 Conclusion

Disease-causing mutations situated within 5′UTRs confirm the importance of motifs in these domains in gene expression and regulation. The ferritin 5′UTR contains a stem-loop structure termed the iron response element, and mutations in this region have been associated with hereditary hyperferritinemia cataract syndrome. It is likely that mutations within the stem-loop alter the structure, resulting in abnormal processing of iron and manifestation of disease (Chatterjee and Pal 2009). Regulation mediated by 5′UTRs involves the combinatorial effects of a multitude of factors and relies heavily on the secondary structure and accessibility of protein binding sites. In addition to the regulatory elements outlined above, it is likely that future investigation will reveal novel factors that interact with the 5′UTR, prior to translation, and influence gene expression.

4 Intronic Regions

Introns are regions of DNA that are transcribed into pre-messenger RNA but are removed during splicing to generate a mature mRNA. Spliceosomal introns are present in all studied eukaryotic organisms. The exact origin of introns is debated, but it is widely accepted that introns evolved soon after the divergence of prokaryotic and eukaryotic organisms and that the current intron content of any particular genome is the result of both intron loss and gain over time (for thoughts and reviews on the topic of intron evolution see (Mattick 1994; Rodriguez-Trelles et al. 2006). Regardless of when and how introns arose, it is clear that the appearance of introns was an important catalyst for evolution, facilitating rapid evolution at the protein level through increased rates of meiotic crossing over within coding regions, as well as rapid evolution of regulatory elements, due to relaxed sequence constraints within non-coding introns (Fedorova and Fedorov 2003). Introns would also have allowed evolution of RNA regulatory pathways without interfering with protein expression, an important distinction that was only made possible by the separation of transcription and translation (Mattick 1994).

4.1 Organization and Length

Intron organization, position and length may influence the ability of the intron to affect gene expression. Intron content varies between different species and some eukaryotic lineages maintain numerous large introns while others seem to have undergone intron loss throughout evolution (Rodriguez-Trelles et al. 2006). The average human gene contains 5–6 introns with an average length of 2,100 nucleotides (Fedorova and Fedorov 2003), although extremes at either end of the spectrum exist. In humans and other animals, intron length is, in general, inversely correlated with transcript levels. A cross-species comparison between yeast, arabidopsis and mouse found that genes involved in stress-response, cell proliferation, differentiation or development generally showed significantly lower intron densities than genes with other functions (Jeffares et al. 2008). Genes in these categories require rapid regulation in response to changing conditions, suggesting that introns may be detrimental to this process. Organisms with short generation times were also found to have a significantly lower genome-wide intron density. Through comparison between the three model organisms, Jeffares and colleagues observed that mouse genes seem to be comparatively less optimised for rapid regulation (i.e. they have higher intron densities), which is logical as mammals are less exposed to rapid environmental changes than plants and microorganisms (Jeffares et al. 2008).

Introns of very different lengths are often found within a gene, although to date, there is no data indicating a global trend concerning length and position, except for the first intron. A large-scale comparison of intron lengths relative to their position in the gene found that the first intron of the CDS tends to be ∼40 % longer than later introns (Bradnam and Korf 2008). Significantly longer first introns were found in species from diverse phylogenetic groups, including vertebrates, insects, plants and fungi, suggesting that this increased length is a common feature of genes in all eukaryotic species. This study also revealed that the first intron was longer again in genes that did not contain an intron within the 5′UTR. In addition to the length of the first intron, a large-scale bioinformatic study that examined 18,217 human ref-sequence genes found these introns, particularly in the first 100 bp, to be enriched for G-rich regions that have the potential to form G4 s (Eddy and Maizels 2008). G4 structures have significant negative effects on translation when located within the 5'UTR of a gene. G-rich elements in the first intron may provide structural targets for regulatory proteins and have an effect on transcription or RNA processing. The position of the first intron relative to the promoter and translation start site means it is a region in which regulatory elements are likely to evolve, as elements within this region are more likely to have a significant effect on promoter activity than elements situated further downstream. In addition, evolution of regulatory elements can occur without disrupting the coding sequence. It is thus likely that the increased relative length of the first intron in many genes is the result of the evolution of regulatory elements (including G4 s) within this region.

4.2 Introns in the UTRs

A genome-wide functional analysis of the 5'UTRs of human genes found that approximately 35 % of human genes contain introns in the 5'UTR (Cenik et al. 2010). 5'UTR introns were found to differ from introns within coding regions with respect to nucleotide composition, length and density, with 5'UTR introns found to be on average twice as long as those in coding regions and generally lower in density. Interestingly, the results from this comprehensive study indicated that the most highly expressed genes tended to have short rather than long 5'UTR introns or lacked them entirely (Cenik et al. 2010). Genes with regulatory roles were also enriched for 5'UTR introns, providing further evidence that the presence of at least one intron within the 5'UTR enhances gene expression either by enhancing transcription or stabilising the mature mRNAs. An intron in the 5'UTR may enhance gene expression through the presence of transcriptional regulatory elements, or through structural modulation and splicing. For example, expression of the ubiquitin C (*UbC*) gene is dependent on the presence of an intron in the 5'UTR. Deletion analyses showed that promoter activity is significantly reduced when the intron is removed, and electrophoretic mobility shift and supershift assays demonstrated that both Sp1 and Sp3 transcription factors bind this region at multiple sites (Bianchi et al. 2009). These experiments indicate that elements within the intron play a major role in the transcriptional regulation of this gene.

In contrast to 5'UTRs, 3'UTRs were found to have relatively few introns (5 %) (Cenik et al. 2010). A study looking at rare cases of intron acquisition in retroposed mammalian genes found that the presence of an intron in the 3'UTR of these genes resulted in down-regulation of gene expression by nonsense-mediated decay (Fablet et al. 2009). This negative effect on expression offers an explanation for the low prevalence of 3'UTR introns. In addition, an in silico study analysing the effect of retained 3'UTR introns upon miRNA target sites indicated that some transcripts only contain miRNA binding sites if the intron in the 3′UTR is retained (Tan et al. 2007). This suggests that variations in intronic splicing in the 3′UTR could result in isoform-specific regulation via miRNAs that may be utilised in a tissue-specific manner.

4.3 Intron Function

Introns could have deleterious effects on gene expression, such as a delay in mature transcript production due to splicing or increased pre-mRNA length, and the energy required to produce a transcript containing introns is also substantially higher. However, the high prevalence of introns in eukaryotic genomes indicates that the benefit must outweigh the potential negative effects. Introns function in a number of different ways and are

- sources of non-coding RNA;
- carriers of transcriptional regulatory elements;
- contributors to alternative splicing;
- enhancers of meiotic crossing over within coding sequences and thus drivers of evolution;
- signals for mRNA export from the nucleus and nonsense-mediated decay (Fedorova and Fedorov 2003).

The effect of introns on genome evolution has already been discussed, but introns also have an important role in the regulation of gene expression, as demonstrated by experiments in which introns are removed or in which introns were inserted into transgenes, resulting in enhanced expression (for an example see Chatterjee et al. 2010). Indeed, many genes with an intact promoter are essentially not expressed at all in the absence of an intron, demonstrating the relative importance of the intronic and promoter regions in some genes (Rose 2008). Introns can enhance gene expression through the presence of transcriptional enhancers or alternative promoters, or by a less well-understood mechanism termed intron-mediated enhancement that arises from introns and increases the processivity of the transcription machinery at the elongation stage. By this mechanism, introns ensure efficient completion of transcription of the gene and could also reduce transcription from sequences that are not genuine promoters (Rose 2008). As well as containing regulatory elements, introns are characterised by a significantly lower nucleosome density in comparison to exons (Nahkuri et al. 2009), and different histone modifications define exons, alternatively spliced exons, and introns (Dhami et al. 2010).

Premature stop codons generally induce degradation of the mRNA via the NMD pathway. Introns seem to act as a signal to differentiate between real and premature stop codons. A study found that splicing of an intron was required to trigger the NMD pathway in PTC-containing T cell receptor-beta mRNA; deletion of the intron abolished NMD (Carter et al. 1996). Further research demonstrated that a large protein complex called the exon-junction complex is deposited just upstream of the junction following splicing, acting as a signal for NMD (reviewed in Chang et al. 2007).

RNA arising from intronic regions may have the ability to regulate the expression of other genes. Polycomb proteins are a group of proteins that act antagonistically to facilitate changes in epigenetic regulation via chromatin remodelling, so coordinated gene expression is required. A recent study demonstrated that an intronic RNA originating from the H3K4 methyltransferase gene, *SMYD3,* binds to EZH2, the core component of the repressive polycomb complex PRC2, and regulates transcription of the corresponding gene. However, overexpression resulted in a decrease in EZH2 transcription and protein levels (Guil et al. 2012). This example shows that an intronic RNA can contribute to complex and coordinated gene expression at the transcriptional level, and importantly can regulate the expression of genes other than the 'parent' transcript. Current thinking is that intronic RNA results from transcription events that are independent from the transcription of the full gene, but it is possible that some may be remnants from mRNA splicing and requiring further investigation.. Antisense non-coding RNA originating from intronic regions has also been reported and is thought to play a role in regulating isoform expression and alternative splicing (Nakaya et al. 2007).

4.4 Regulatory Elements: Enhancers

Enhancers are segments of DNA that enhance transcription of genes by interactions with *trans*-acting factors. Enhancers generally interact in a specific manner with the corresponding promoter through chromatin looping of the intervening DNA, to associate enhancer-bound transcription factors with the promoter (Nolis et al. 2009), and recent data have indicated that enhancers may also affect downstream processes, such as decompaction of the chromatin fibre and the release of RNAPII (Ong and Corces 2011). Although these elements interact specifically with the promoter, enhancers are variable, and upstream, downstream and distal elements have been identified that can activate transcription, independent of their location or orientation with respect to the promoter (Ong and Corces 2011). Enhancers are now recognised as the main regulatory elements involved in transcription and many enhancer elements are critical in defining the expression patterns of genes. An enhancer element situated within an AT-rich regulatory region in the first intron of *Imp2* is critical for the expression of this gene. This enhancer serves as a binding site for HMGA2 that acts to recruit and stabilise a complex of transcription factors, resulting in *Imp2* transcription (Cleynen et al. 2007). Mutations that disrupt enhancer activity may also have a profound effect on the expression of the downstream gene. Enhancer activity in the *OCA2* gene is strongly associated with variation in human eye colour (Duffy et al. 2007). SNPs disrupting a conserved enhancer that binds helicase-like transcription factor

(HLTF) upstream of this gene reduce the expression and result in blue eye colour, with a frequency of 78 % (Sturm et al. 2008). This emphasizes the importance of many enhancers in regulating gene expression and provides evidence that variations within enhancers are likely to contribute to individual phenotype and disease susceptibility.

Recent studies using genome-wide tools have indicated that many enhancers are associated with specific histone modifications, that allow them to be recognised and utilised in a specific manner (Ong and Corces 2011). Promoters can generally be influenced by distinct enhancer elements under varying conditions (Maston et al. 2006), while binding of factors that do not associate strongly with the promoter may "switch off" the enhancer as required. An enhancer region that is critical for specific gene expression during development is the human-accelerated conserved non-coding sequence 1 (*HACNS1*). This element is the most rapidly evolving human non-coding element identified to date and experiments using a transgenic mouse model showed that this element drove strong and specific reporter gene expression in the anterior limb bud, pharyngeal arches, and developing ear and eye, indicating that *HACNS1* acts as a robust enhancer during development (Prabhakar et al. 2008). In contrast, the chimpanzee orthologue failed to drive reproducible reporter gene expression in a similar manner, suggesting that this region is vital for development of human-specific digit and limb patterning that distinguishes humans from other primates, specifically bipedialism and dexterity of the human hand.

The complexity arising from enhancers is increased by the fact that often multiple enhancers and other elements interact and have a combinatorial effect on gene expression. The cystic fibrosis transmembrane conductance regulator (*CFTR*) gene is activated by coordinated regulation from several intronic enhancers that bind both tissue-specific and general transcription factors (Ott et al. 2009). Differential interactions between the various enhancers and the promoter were found to result in variable expression levels in epithelial cells of intestinal lineage (high expression) and of the respiratory system (lower expression) and chromatin conformation capture was used to identify distal regulatory sites that also contributed to gene expression. This illustrates how complex interactions between enhancers and distal elements can contribute to the tissue-specific expression of a gene. In addition to controlling the differential expression of a single gene, conserved enhancers contribute to the regulation of whole gene pathways. Transcription factor Ronin and the transcriptional coregulator Hcf-1 are essential factors involved in the self-renewal of embryonic stem (ES) cells. They bind to a highly conserved enhancer element in a subset of genes that function in transcription initiation, mRNA splicing and cell metabolism (Dejosez et al. 2010). The enhancers that bind Ronin/Hcf-1 are thus key elements required for ES cell pluripotency.

In vivo analyses of evolutionarily conserved non-coding sequences revealed an enrichment of developmentally specific *cis*-regulatory transcriptional enhancers (Prabhakar et al. 2008). Indeed, the high proportion of non-coding to coding regions in the human genome compared to other species provides strong evidence that the complexity of humans arises from evolution of these non-coding regions, with enhancers likely to play a major role in this process.

5 3′ Untranslated Region

The 3′ untranslated region (3′UTR), situated downstream of the protein coding sequence, is involved in numerous regulatory processes, including transcript cleavage, stability and polyadenylation, translation and mRNA localisation. They are thus critical in determining the fate of an mRNA. In comparison to the 5′UTR, which contains sequences responsible for translation initiation, sequence constraints within the 3′UTR are more relaxed resulting in a greater potential for evolution of regulatory elements. Despite this, regions of high conservation are also prevalent, with 3′UTRs containing some of the most conserved elements within the mammalian genome (Siepel et al. 2005). A genome-wide in silico analysis revealed that contrary to the promoter region, motifs in the 3′UTR are primarily conserved on one strand, which is consistent with the 3′UTR acting to regulate gene expression at the post-transcriptional level (Xie et al. 2005). The 3′UTR serves as a binding site for numerous regulatory proteins as well as microRNAs (Fig. 1c) and in order to understand the properties of this region it is necessary to first discuss the research history of these interactions.

5.1 MicroRNAs and the 3′UTR

MicroRNAs (miRNAs) are endogenous, single-stranded non-coding RNA molecules of approximately 22nt in length that interact with mRNA targets post-transcriptionally to regulate expression. In animals, miRNAs generally exert an effect by partial base pairing to a miRNA response element (MRE) on a target mRNA via a 'seed sequence' at the 5′ end of the miRNA, which then recruits Argonaut and inhibits translation of the mRNA (see Song et al. 2008; Paik et al. 2011; Gerin et al. 2010). Another mechanism by which miRNAs can down-regulate genes is through perfect base pairing with a target sequence, promoting RNA cleavage as a result of incorporation into the RNA-induced silencing complex (MacDonald et al. 1993), although only a few examples of this have been described (Yekta et al. 2004). In addition to down-regulating gene expression, some miRNAs have been found to induce translational up-regulation; validated targets of this type of regulation include the tumour necrosis factor-alpha and the cytoplasmic beta-actin gene (Vasudevan et al. 2007; Ghosh et al. 2008). Data indicates that miRNA repression occurs in proliferating cells, while activation is mediated by some miRNAs during cell cycle arrest (Mortensen et al. 2011; Vasudevan et al. 2007). miRNAs are the most extensively studied group of non-coding RNAs and interested readers are referred to current reviews on miRNA functions and mechanisms (Huang et al. 2011; Jeffries et al. 2009; Fabian et al. 2010), miRNA response element prediction (Saito and Saetrom 2010), miRNA-mediated regulation of developmental processes (Williams et al. 2009; Zhao and Srivastava 2007), regulation of miRNA expression (Krol et al. 2010) and the impact of miRNAs on evolution of 3′UTRs (Zhang and Su 2009).

A wealth of information regarding miRNA expression and function is now available and it is evident that miRNAs are a vital component of gene control. miRNAs have been found to be involved in most important biological events, including cell proliferation and differentiation, development, nervous system regulation and tumourigenesis (reviewed in Huang et al. 2011), and common miRNA targets include transcription factors and signalling proteins (Zhang and Su 2009). An individual miRNA has the ability to regulate a large number of target genes because complementarity is only required in the seed region, and miRNAs may be involved in the regulation of a process or system. For example, muscle growth and differentiation is modulated by a set of miRNAs that are themselves controlled by myogenic transcription factors (Williams et al. 2009). The cardio-vascular system is also controlled by a specific set of miRNAs, which influence processes such as cell cycle progression, cardiomyocyte differentiation, and vessel formation (Cordes and Srivastava 2009). In accordance with this, dysregulation of miRNAs can have major impacts on these processes and has been associated with cardiac rhythm abnormalities and hyperplasia (Zhao et al. 2007), coronary artery disease (Fichtlscherer et al. 2010), myocardial infarction (Fichtlscherer et al. 2010; Bostjancic et al. 2010), and numerous muscular diseases including Huntington's disease and muscular dystrophy (Eisenberg et al. 2009). These are just two examples of systems in which a group of miRNAs act as a regulatory layer to fine-tune gene expression, enabling the system to operate efficiently through the coordinated expression of the associated genes.

An mRNA may be regulated by multiple different miRNAs, expanding the repertoire of expression of an mRNA at a given time, in a particular cell type. Studies on MRE prediction and validation have shown that the presence of multiple seed sequences within an mRNA is common ($\sim$50 % of targets) and targets are frequently expressed in a mutually exclusive manner to the miRNA, further indicating a role for miRNAs in fine-tuning of gene expression and developmental processes (Stark et al. 2005). As miRNAs often have a subtle effect on gene expression (e.g. twofold down-regulation), the combination of multiple miRNAs acting at once could invoke a much stronger repression. The chain of events following miRNA interaction with a target can influence additional miRNA binding, so the presence of multiple seed sequences could also be a safeguard to ensure expression of the target is tightly controlled, even if levels of the various miRNAs fluctuate. miRNAs may also interact with various RNA binding proteins. The dead end 1 RBP binds a target sequence and counteracts the function of a number of miRNAs (Kedde et al. 2007). These types of interactions, whereby gene expression is modulated according to the concentrations of various RNAs and factors through a cascade of interacting events, seems to be the overriding theme of the eukaryotic regulatory system. It is no surprise then that mutations that change the expression of miRNAs can have dire effects in the organism. Trisomy 21, the cause of Down syndrome, has a severe and complex phenotype. *In silico* analysis has shown that five miRNA genes are duplicated in this event, and overexpression of these genes has been proposed to reduce the expression of target genes, contributing to the severe phenotype of this syndrome (Elton et al. 2010).

Many miRNAs are evolutionarily conserved (Zhao and Srivastava 2007; Bartel 2004) and the lack of requirement for long regions of complementarity means that novel miRNAs and MREs can easily arise, implicating them as powerful tools for evolution (Stark et al. 2005). miRNAs bind preferentially in the 3′UTRs of protein-coding genes, although some target sites have been identified in the 5′UTR and intronic gene regions. An inter-species genome-wide comparison found that motifs in the 3′UTR are an average of 8 bp in length and that around half of all the motifs identified are likely to be related to miRNAs (Xie et al. 2005). miRNAs are often expressed in a tissue-specific or developmental stage-specific manner and genes involved in processes common to all cells have evolved to selectively avoid sequences complementary to miRNA seed regions (Stark et al. 2005). This mechanism of selective avoidance has a significant impact on the evolution of the 3′UTR. A recent study found that modification of the stop codon to extend the coding region of a transgene reporter changed the mechanism from miRNA-induced translational repression to RISC-mediated degradation by small interfering RNAs (Gu et al. 2009). These results indicate that active translation impedes miRNA-RISC interaction with target mRNAs and provides an explanation as to why MREs are contained in the non-coding regions. Data obtained *in vitro* and *in vivo* supported the conclusion that while siRNA can work efficiently in non-coding and coding regions, miRNA activity is significantly inhibited when targeting the coding region, indicating that miRNA-programmed RISC is required to remain attached to the target mRNA to effectively silence translation *in cis* (Gu et al. 2009). Data also provided a possible explanation for the low prevalence of MREs situated in the 5′UTR, as scanning of the 5′UTR by the translation initiation complex may impair formation of miRNA-RISC complexes.

miRNAs, like most other regulatory molecules, are part of a larger system in which numerous factors impact on the ultimate expression of the target gene. miRNA expression can also be altered by environmental factors, in a study on miRNA expression in primary murine macrophages, following inflammatory response, one miRNA (miR-155) that was significantly upregulated was identified (O'Connell et al. 2007). The inflammatory response involves the upregulation of a large number of genes, and miR-155 is a common target that acts to control the expression of many of the genes during the process, preventing overexpression and reducing the risk of cancer. Other environmental factors including diet, alcohol intake, stress and exposure to infectious agents and carcinogens also impact miRNA expression (as well as other epigenetic modifications) (Mathers et al. 2010). It is likely that miRNAs and other regulatory mechanisms that involve the coordinated expression of many genes and interacting factors are responsible for the phenotypic variations we see within eukaryotic species.

5.2 Stabilisation and AU-Rich Elements

Modification of transcript stability allows expression to be rapidly controlled without altering translation rates. This mechanism has been found to be critically involved in vital processes such as cell growth and differentiation, as well as adaptation to external stimuli (Eberhardt et al. 2007; Elkon et al. 2010). The most well characterised stabilisation elements are the AU-rich elements (Jeffares et al. 2008) situated in the 3'UTR of some genes. These elements range in size from 50 to 150 bp and generally contain multiple copies of the pentanucleotide AUUUA (Chen and Shyu 1995). AREs play a critical role in the stability of particular genes. Early studies indicated that AREs are variable in sequence and three main classes have been defined that differ in the number and arrangement of motifs. Class I AREs contain one to three scattered AUUUA motifs within a U-rich region, while class II AREs are characterized by multiple overlapping AUUUA motifs. The third class of AREs lack the AUUUA motif but contain U-rich stretches (Chen and Shyu 1995). AREs bind proteins (ARE-BPs) that generally promote the decay of the mRNA in response to a variety of intra- and extra-cellular signals (for some recent examples see (Chamboredon et al. 2011; Knapinska et al. 2011; LaJevic et al. 2010), although binding proteins that act to regulate translation have also been described (Lopez De Silanes et al. 2007). Genes regulated by AREs include cytokines, growth factors, tumour suppressors and proto-oncogenes, as well as genes involved in the regulation of the cell cycle, such as cyclins, enzymes, transcription factors, receptors and membrane proteins (Eberhardt et al. 2007). This plethora of vital gene families carrying these elements affirms the significance of transcript stability in the process of gene regulation.

Many ARE-BPs are expressed in a tissue- or cell-type specific manner (Reznik and Lykke-Andersen 2010), with ARE secondary structure being an important factor in ARE-BP activity (Meisner et al. 2004). Different ARE-BPs can compete for the same binding site and depending on the cellular localisation, environment and timing, regulation from an ARE can result in different outcomes for a transcript. A class III ARE in the *c-jun* 3'UTR has been shown to decrease steady-state mRNA levels but also be involved in increasing protein production (Barreau et al. 2006). This seems counterintuitive, but it is likely that each mechanism is used at different times for different needs, such as in developmentally or tissue-specific circumstances. Environmental factors can also impact ARE protein binding, with stability playing a major role in response to stresses such as heat shock and nutrient deprivation. These stimuli trigger a signalling cascade that alters the abundance of various ARE binding proteins, while simultaneously manipulating RNA binding properties (reviewed in Eberhardt et al. 2007). Expression of the anti-apoptotic protein $Bcl-X_L$ is increased by stabilisation following UVA irradiation, a process implicated in skin and other cancers. Examination of the ARE-BPs associated with an ARE in the $Bcl-X_L$ 3'UTR identified nucleolin as a key stabilising protein and the authors suggest that UVA irradiation increases the binding capacity of nucleolin to the ARE and facilitates protection of the $Bcl-X_L$ mRNA from degradation (Zhang et al. 2008).

In addition to affecting stability, AREs have also been found to activate translation, although this pathway is less common and is poorly understood. The 3′UTR of cytokine tumour necrosis factor α (*TNFα*) mRNA contains a highly conserved 34nt ARE (Vasudevan and Steitz 2007). This gene is expressed in stimulated lymphocytes and is critical for inflammatory response so must be rapidly regulated when required. During inflammation, cell growth is arrested and up-regulation of TNFα occurs at the protein level. Studies found that Argonaut 2 (AGO2) and fragile-X mental retardation syndrome-related protein 1 (FXR1) associate with the ARE of *TNFα* and function to activate translation in response to serum starvation (Vasudevan and Steitz 2007). It was also found that human miR369-3 binds through the seed sequence to the ARE and directs association of these factors with the ARE to activate translation, providing evidence for a secondary role of miRNAs in translation, alongside their well-studied destabilising roles (Vasudevan et al. 2007). An earlier study examining the structure of the *TNFα* ARE showed that hairpin folding modulates binding of proteins to that motif and mediates different outcomes for the mRNA (Fialcowitz et al. 2005). These experiments demonstrate the versatility of AREs, RNA-binding proteins and miRNAs in modulating gene expression in a positive or negative manner, as required. The ability of AREs to influence both mRNA stability and translation is likely to result from different signals. The GU-rich element (GRE) is another recently discovered stability element that interacts with CUGBP1, an RNA binding protein that promotes decay of the associated mRNA (Lee et al. 2010; Vlasova et al. 2008). Alongside microRNAs, AREs and GREs have impacted upon the evolution of the 3′UTR, and thus shaped the regulation of gene expression from this region.

5.3 Structure

5.3.1 Poly(A) Tail

The poly(A) tail results from the addition of a series of adenosine bases to the 3′ end of an RNA transcript. This provides the mRNA with a binding site for a class of regulatory factors called the poly(A) binding proteins (PABP) that have roles in the regulation of gene expression, including mRNA export, stability and decay, and translation (reviewed in Mangus et al. 2003; Gorgoni and Gray 2004; Goss and Kleiman 2013), playing vital roles during vertebrate development (Gorgoni et al. 2011). Five different PABPs have been identified in humans, one nuclear and four cytoplasmic, all of which have distinct functional roles (Gorgoni et al. 2011). Further research will undoubtedly characterize each PABP but it seems that while there are some common functions for all the PABPs, subtle differences exist and are evident through interactions with proteins and RNA (reviewed in Goss and Kleiman 2013). PABPs seem to function as scaffolds for the binding of numerous other factors, thus they indirectly regulate gene expression. Aside from their global effects on translation, PABPs can also regulate the translation of individual mRNAs, although this is less well documented (e.g. Cyclin B, Cao and Richter 2002). A recent study demonstrated that PABPC1 binding in the 3'UTR of the *YB-1* mRNA stimulates translation in a specific manner (Lyabin et al. 2011). However, YB-1 also binds in the same region and inhibits translation, so competitive binding with PABPC1 determines whether expression is up- or down-regulated. PABP mRNAs can also bind poly(A) tracts in their own 5'UTRs, repressing their own translation and maintaining balance and controlled regulation. The poly(A) tail is synthesised at a defined length (~ 250 bp in mammalian cells), which may then be shortened in the cytoplasm to promote translational repression as required (Kuhn et al. 2009).

5.3.2 5'–3' Interactions

Early experiments investigating the roles of the 5'cap structure and the poly-A tail found that they function synergistically to control mRNA translation (Gallie 1991). The addition of a poly(A) tail to a luciferase reporter gene increased protein expression 97-fold when the length of the 3'UTR was 19 bases (Tanguay and Gallie 1996), demonstrating the essential role of the poly(A) tail in efficient translation. The association of PABPs with the poly (A) tail facilitates an interaction with eIF4F bound to the 5'cap structure, resulting in circularisation of the mRNA that promotes translation initiation and ensures ribosome recycling and efficient translation (For reviews on translation initiation and the 5'–3' interaction pathway see (Jackson et al. 2010; Chen and Kastan 2010; Mazumder et al. 2003). This interaction also allows inhibition of translation by inhibitor proteins bound to the 3'UTR, which is important because the relative lack of constraint in RNA secondary structure in the 3'UTR compared to the 5'UTR indicates that response to changing conditions can occur with fewer consequences while feeding back information to the initiation site (Mazumder et al. 2003). In addition to binding through protein interactions at the 5'cap structure, sequence specific interactions between the 5' and 3' ends of an mRNA have also been observed. The human p53 gene contains a region of complementarity between the 5' and 3'UTRs that have been shown to interact and bind translation factor RPL26 that mediates translational up-regulation as a response to DNA damage (Chen and Kastan 2010). Because of the importance of the 5'–3' interaction pathway, mutations affecting the termination codon, poly-adenylation signal and secondary structure of a 3'UTR can cause translation de-regulation and disease (Chatterjee and Pal 2009).

A genome-wide analysis of UTRs identified numerous motifs within human 5'UTRs that were specific to the 3' ends of miRNAs, with many of these found to simultaneously contain 5' end interaction sites in the 3'UTRs (Lee et al. 2009). Further investigation demonstrated that interactions between the 5' and 3' ends of many genes are facilitated by an interaction with a single miRNA, and that genes highly influenced by miRNA overexpression or deletion contained predicted binding sites in both UTRs. The authors termed this class of miRNA targets miBridge, and reporter gene assays revealed that deletion of either binding site reduced repression from the miRNAs, indicating that the interaction is essential for potent down-regulation of the transcript (Lee et al. 2009). It is clear that interactions between the 5′ and 3′UTR contribute to the precise control of expression pathways and responses, and mRNA circularisation provides an explanation as to how translation can be so efficiently repressed via protein or miRNA binding in the 3′UTR.

5.3.3 Length

The requirement of 5′–3′ interactions for efficient translation has implications for both the length and secondary structure of the 3′UTR, with studies demonstrating the significant impact of some longer 3′UTRs on expression. Using a luciferase reporter gene, Tanguay and Gallie (1996) observed that increasing the length of the 3′UTR from 19nt to 156 nt decreased expression ∼45-fold, independently of the orientation, gene or sequence (Tanguay and Gallie 1996). This early example indicates 3′UTR length is a major determinant in mRNA expression. Aside from the importance of interaction with the 5′UTR, the prevalence of miRNA binding sites also has an impact on the length, as longer 3′UTRs are more likely to possess miRNA binding sites that have the potential to inhibit translation. A study comparing the length and miRNA-binding site content of ribosomal and neurogenesis genes found that ribosomal genes had shorter 3′UTRs and specifically devoid of miRNA-binding sites, when compared to random controls (Stark et al. 2005). In contrast, 3′UTRs of genes involved in neurogenesis were longer and specifically enriched for potential binding sites. The *Hip2* gene uses alternative 3′UTRs to control expression as required. The longer 3′UTR of this gene contains conserved seed matches to two miRNAs that are expressed in activated T-cells (Sandberg et al. 2008). Upon activation, relative expression of the transcript with the longer 3′UTR decreased and protein expression significantly increased. This is consistent with a model in which use of alternative 3′UTRs prevents down-regulation by miRNAs, allowing up-regulation of protein production.

In general, longer 3′UTRs correlate with a relatively lower expression level, as indicated by experiments comparing the expression of isoforms differing only in their 3′UTR (Sandberg et al. 2008). The *SLC7A1* gene is expressed with two variant 3′UTRs, the longer of which contains an additional potential miRNA binding site. A functional polymorphism in this gene has been associated with endothelial dysfunction and genetic predisposition to essential hypertension. However, this allele was found to be preferentially associated with the longer 3′UTR, resulting in decreased expression compared to the wild-type allele (Yang and Kaye 2009). Notably, the average length of the 3′UTR in humans is more than twice that of other mammals (Pesole et al. 2001), which is indicative of an increase in regulatory elements in human genes. Although it is clear that miRNAs impact on 3′UTR length, other factors are also likely to contribute, potentially in a developmentally or tissue-specific manner. The relative position of motifs such as

AREs within the 3'UTR can affect protein binding and regulation. The β_2-adrenergic receptor (β_2-AR) 3'UTR contains a number of AREs, although translational suppression seems to be primarily mediated by a 20nt ARE and a poly(U) region situated at the distal end of the 3'UTR. These motifs have been shown to bind T-cell-restricted intercellular antigen-related protein (TIAR) that acts to repress translation, and HuR, an ARE-BP that can stabilise transcripts (Kandasamy et al. 2005). Recent experiments using reporter constructs demonstrated that the length of the 3'UTR is critical for these interactions, as TIAR binding was reduced in constructs with a shorter 3'UTR ($\sim$100nt) in comparison to constructs with longer 3'UTRs (300 and 500nt) (Subramaniam et al. 2011). HuR binding was not affected, indicating the two factors bind at non-overlapping sites and exert different controls on expression, increasing the complexity of regulation of this gene.

5.3.4 Secondary Structure

Secondary structures within the 3'UTR are emerging as more important than previously envisioned. While the length of the 3'UTR is important, the secondary structure folding is also a vital determinant of translation efficiency and mutations that change the secondary structure may result in disruption of expression. A study by Chen et al. (2006) on 83 disease-associated variants in the 3'UTR of various human mRNAs found a correlation between the functionality of the variants and changes in the predicted secondary structure (Chen et al. 2006). NMD is a quality control mechanism to remove mutated non-functional transcripts. Most commonly, the location of the nonsense mutation relative to the exon–exon junction complex determines the efficiency of NMD (Chang et al. 2007), but the 3'UTR may also play a role. The mechanisms of translation termination at premature termination codons (PTCs) has been shown to rely on the physical distance between the termination codon and the poly-A binding protein, PABPC1 (Eberle et al. 2008). This study found that extending the region between the normal termination codon and the poly-A tail resulted in NMD and that spatial rearrangements of the 3'UTR can modulate the NMD pathway (Eberle et al. 2008).

Secondary structure of the 3'UTR is difficult to predict because of the multitude of factors binding the region, many of which are likely to induce structural changes. Factors can changes the spatial configuration of the region by disrupting mRNA folding, or by interacting with other factors resulting in the looping out of the mRNA in between (Eberle et al. 2008). The stem-loop RNA structure is the most common example of a secondary structure that can modify gene expression, and in the 3'UTR this generally occurs through RNA-binding proteins. Brain-derived neurotrophic factor transcript (*BDNF*) contains an extended stem-loop structure that is responsible for the stability of the mRNA in neurons in response to Ca^{+2} signals (Fukuchi and Tsuda 2010). The authors suggest that the stem-loop structure provides a scaffold for the interaction of a number of RNA binding proteins, non-coding RNAs and poly-adenylation factors in response to Ca^{+2}.

In *TNFα*, an ARE in the 3′UTR adopts a stem-loop structure that has been shown to modulate its affinity for various ARE-BPs (Fialcowitz et al. 2005). These examples demonstrate that modulation of 3′UTR secondary structure by protein binding or other means can modulate *trans*-factor binding specificity and thus contributes to gene regulation at the post-transcriptional level.

5.3.5 Alternative 3′UTRs

Alternative poly-adenylation (APA) and alternative splicing are two mechanisms that can result in the production of mRNA isoforms differing in their 3′UTR. APA can occur because of the presence of multiple poly-adenylation sites, or by mutually exclusive terminal exons, and it is estimated that APA is utilised by ∼50 % of human genes (Dickson and Wilusz 2010). These mechanisms are very useful for complex organisms, as they provide a way in which transcripts can express the same protein but with varying expression levels and/or spatial localisation arising from variation in regulation from the 3′UTR (Sandberg et al. 2008). Alternative 3′UTR use is an important aspect of developmental- and tissue-specific gene expression (Ji et al. 2009; Wang et al. 2008; Hughes 2006; Ji and Tian 2009) (for an example see (Winter et al. 2007) and large-scale changes in APA patterns have been associated with a number of different cancers (Mayr and Bartel 2009; Fu et al. 2011). APA also plays an important role in isoform localisation (Andreassi and Riccio 2009). The *HuR* gene is an ARE-BP that is involved in the stabilisation of many ARE-containing mRNAs. APA produces a number of *HuR* variants that differ in expression levels, and while the predominant transcript lacks AREs, a rare variant has been identified that contains functional AREs in the 3′UTR (Al-Ahmadi et al. 2009). These AREs were found to bind HuR, thus inducing a self-upregulation loop. Use of alternative 3′UTRs allows versatility of expression from a single gene.

5.4 Conclusions

The 3′UTR is a versatile region that is enriched for regulatory elements and is vital for correct spatial and temporal gene expression. The 3′UTR is also emerging as a major hotspot for interactions with non-coding RNAs, with recent studies showing that a large number of 3′UTRs are also expressed independently from the primary gene transcript and are likely to function *in trans* as non-coding RNAs of various lengths (Mercer et al. 2010). Further investigation into the regulatory functions of 3′UTRs has the potential to reveal even more complex pathways and interactions.

6 Non-Coding RNAs

Over the past decade, a wealth of evidence has revealed the pervasiveness and complexity of transcription throughout the human genome, with the majority of bases associated with at least one primary transcript (Birney et al. 2007). As less than 1.5 % of the human genome codes for protein, this process results in widespread production of non-coding RNAs, of which there are many different types (interested readers are referred to reviews for each category), including miRNAs (Saito and Saetrom 2010; Jeffries et al. 2009; Zhang and Su 2009; Williams et al. 2009), promoter-associated RNAs (Preker et al. 2008; Fejes-Toth et al. 2009), short interfering RNAs (Okamura and Lai 2008; Watanabe et al. 2008), piwi-interacting RNAs (Klattenhoff and Theurkauf 2008; Lin 2007), small nuclear RNAs (Dieci et al. 2009), natural antisense transcripts (Su et al. 2010; Faghihi and Wahlestedt 2009) and long non-coding RNAs (Ponting et al. 2009; Clark and Mattick 2011; Mercer et al. 2009; Wilusz et al. 2009), long intronic non-coding RNAs (Louro et al. 2009), and RNAs as extracellular signalling molecules (Dinger et al. 2008). Non-coding RNAs can be sense or antisense in orientation, transcribed in either direction and can originate from intergenic and intronic regions. Although there are some examples of non-coding RNAs conserved between distant species (Wahlestedt 2006), the majority of non-coding RNAs seem to be species-specific, at least at the sequence level (Hawkins and Morris 2008). However, recent studies have shown that thousands of sequences within the mammalian genome possess conserved RNA secondary structures, while lacking any significant sequence conservation (Torarinsson et al. 2006, 2008). Some non-coding RNAs are likely to function primarily through their secondary structures, which would result in relaxed sequence constraints and an underestimation of conservation between species. In any case, it is apparent that contrary to previous assumptions, a lack of conservation is not necessarily indicative of a non-functional sequence and genome-wide evidence indicates that a significant proportion of non-coding RNAs perform functional roles (Mercer et al. 2009).

Non-coding RNAs are key regulators of gene expression, acting at the individual gene level, regulating *cis* and *trans* interactions and contributing to control of transcription and translation, and on a genome wide-scale, regulating accessibility of chromatin and controlling gene pathways. Non-coding RNAs associate with each of the untranslated gene regions discussed in this review, contributing to the fine control of gene expression and increasing the complexity of the regulatory system. Transcribed regions including the 5′ and 3′UTRs and intronic regions are also likely origins of non-coding RNA, following splicing and translation of the associated gene (Mercer et al. 2010). The use of RNA as a regulatory element has advantages because it can rapidly be synthesised and degraded (Djupedal and Ekwall 2009), has structural plasticity and can modulate gene expression in response to external factors (Ansari 2009) and can act combinatorially to control complex interactions and regulatory pathways (Mattick 2004). The discovery of non-coding RNAs, which were largely unnoticed previously, has come about due

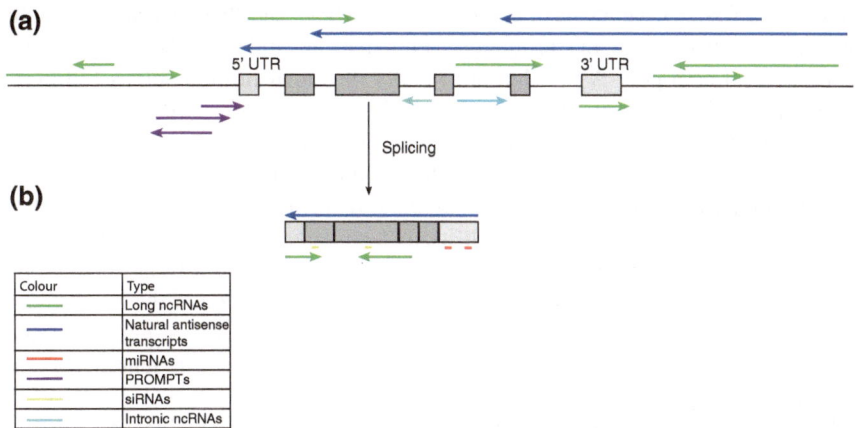

Fig. 2 Pervasive expression of non-coding RNAs in the eukaryotic genome. **a** Non-coding RNA expression relative to a typical gene. Exons are represented by the *grey boxes*, the *light-grey boxes* represent the untranslated regions. Non-coding RNAs are expressed from all regions of the genome, in varying sizes and transcribed in either direction (see Table). Long non-coding RNAs and natural antisense transcripts can target DNA, pre-mRNA or **b** mRNA. Non-coding RNAs may themselves be spliced; this is the case for many natural antisense transcripts that are complementary to their target gene. Smaller regulatory RNAs such as miRNAs and siRNAs target mRNA to regulate gene expression post-transcriptionally. Long ncRNAs are the most widespread category, as the other types are defined by a specific origin, localisation, or direction

to advances in detection methods and technologies. Non-coding RNAs have now been identified to originate from all regions of the genome, although they are more frequently derived from regions surrounding gene promoters, enhancers and 3'UTRs (He et al. 2008). This is indicative of a key role in the control of translation and stability. An *in vitro* study examining five different human cell types showed that the distribution of non-coding RNAs was non-random across the genome, differed among cell types, and that the distribution of sense and antisense transcripts were distinct (He et al. 2008). In particular, antisense transcripts were concentrated around gene promoters and 3'UTRs, while sense transcripts were more prevalent around exons. Non-coding RNAs have now been found to control all aspects of gene expression. The following discussion will explore each type of non-coding RNA and illustrate the versatile nature of non-coding RNA control of gene expression, with the exception of miRNAs that have already been covered. See Fig. 2 for a visual summary of the different types of non-coding RNA.

6.1 Promoter Associated RNAs (PROMPTs)

PROMPTs (Promoter upstream transcripts) are non-coding RNAs originating from 500 to 2,500nt upstream of active gene promoters. It is thought that most actively transcribed RNAPII genes have associated PROMPTs and that they are especially prevalent at dispersed CpG promoters (Preker et al. 2008). PROMTs possess a 5'cap as well as a poly-A tail, but are rapidly degraded in the nucleus following transcription (Preker et al. 2011), a characteristic that is indicative of a regulatory role. 5' RACE analysis and evidence that PROMPTs are transcribed in both directions suggests that they rely on the promoter of the associated gene for transcription rather than a promoter of their own. The function of PROMPTs currently remains unknown, although it has been suggested that they may act as transcription enhancers, because the occupation of RNAPII upstream of the promoter may alter the chromatin landscape to ensure transcription of the downstream gene (Preker et al. 2011). These transcripts are also likely to play a regulatory role, especially considering they are generally hundreds of bases in length and are capped and polyadenylated. It will be interesting to see what transpires in this field, as the role(s) of these articular non-coding are further elucidated.

6.2 Short Interfering RNA (siRNA)

siRNAs are short, double stranded non-coding RNAs of 20–25 bp that were first discovered in plants (Hamilton and Baulcombe 1999). siRNAs bind complementary sequences in mRNAs and induce degradation by the RNA interference pathway, in which the targeting strand is incorporated into RISC and results in cleavage by Argonaute and post-transcriptional gene silencing. Gene regulation by siRNA seemed to only occur in organisms with RNA-dependent RNA polymerase (RdRP), a polymerase not found in mammals, but more recent studies have shown siRNAs are present in mammals, albeit in low abundance (Watanabe et al. 2008). This study showed that siRNAs are present in mouse oocytes and that they regulate complementary genes via the RNAi pathway. When present, it seems that siRNAs play a small role in the non-coding regulatory system of higher eukaryotes (although with the advent of deep sequencing this could change). However, soon after their discovery it was demonstrated that synthetic siRNAs could be used to selectively knockdown gene expression in mammalian cells (Elbashir et al. 2001), and they have since been widely used to investigate gene function and are being exploited to develop disease therapeutics.

6.3 Piwi-Interacting RNAs (piRNAs)

piRNAs are a class of non-coding RNAs, 29–30nt in length, that specifically interact with Piwi proteins (Girard et al. 2006). Piwi proteins are a class of Argonaute proteins expressed in germline cells and are involved in cell division and stem cell maintenance, but their exact biological mechanism remains to be elucidated (Carmell et al. 2002). Because of this, the function of piRNAs also remains to be determined, but the abundance of piRNAs in germline cells and the demonstration of male sterility in mice with targeted mutations to the Piwi proteins suggests they play a role in gametogenesis (Girard et al. 2006).

6.4 Small Nuclear RNA (snRNA)

This class of noncoding RNAs are found in the nucleus of eukaryotic cells and are around 150nt in length. These were one of the first types of non-coding RNA to be identified, and the reason for this is because they are involved in splicing. snRNAs are a component of the small nuclear ribonuclear protein complexes (snRNPs) that make up the spliceosome, binding to complementary regions in precursor RNA and facilitating assembly of the spliceosome (Wassarman and Steitz 1992). Five different snRNPs are invovled in splicing, U1, U2, U4, U5 and U6. During splicing they interact with different parts of the pre-mRNA; U1 binds the conserved 5′ splice site, U2 binds the conserved branch point consensus sequence, and U4-6 bind together to form a complex before completing the spliceosome through interaction with the other snRNAs (Wassarman and Steitz 1992). snRNAs evidently play a vital role in the eukaryotic genome, acting in a very specific manner to control the global processing of RNA. Some other examples of "housekeeping" non-coding RNAs are ribosomal RNA, which is the RNA component of the ribosome, and transfer RNAs, which mediate peptide synthesis. Housekeeping non-coding RNAs play an essential role in gene processing by actively interacting with proteins and enabling correct localisation and function.

6.5 Cis-Natural Antisense Transcripts

These are transcripts that are complementary to expressed genes and function to regulate their expression (Faghihi and Wahlestedt 2009). In general, NATs appear to have a repressive effect on the expression of the corresponding gene, acting to ensure expression is maintained at the required level. For example, BDNF expression is repressed by a conserved NAT that overlaps with 225nt of BDNF at the 3′end (Modarresi et al. 2012). This NAT initiates ∼200 kb downstream of the BDNF promoter and undergoes splicing. Knockout of this transcript resulted in a two to sevenfold up-regulation of BDNF protein over RNA (Modarresi et al. 2012). There are a number of proposed mechanisms for NAT-directed regulation of sense mRNA (Faghihi and Wahlestedt 2009). One suggestion involves "transcriptional collision", in which the transcriptional machinery collides during transcription of each strand, due to the opposing orientation of the mRNA and NAT transcription. However, transcription can occur at different times and this unlikely to be the major mechanism of regulation. The observation that NATs are often associated with imprinted genes also suggests some NATs may interact with

DNA and induce epigenetic changes to completely repress transcription. In addition, the formation of an RNA–RNA duplex, either in the nucleus or cytoplasm, can affect the sense transcript by altering localisation, stability, or translational efficiency. NATs are apparently common non-coding RNAs associated with genes, and future research is likely to reveal additional details on the function of such transcripts.

6.6 Long Non-coding RNAs

A large and varied class of transcripts, termed long non-coding RNAs, are defined as transcripts greater than 200nt in length. As such, they have many different functions and characteristics, many of which were discussed earlier in this review. Long non-coding RNAs can originate in the coding and intergenic regions, and are widely transcribed throughout the genome. An example of a long non-coding RNA is *HOTAIR,* the expression of which increased in breast cancer (Gupta et al. 2010). *HOTAIR* is localised at the *HOX* locus, which contains *HOX* genes that code for transcription factors involved in segmental differentiation during embryo development. Elevated expression of *HOTAIR* induced chromatin remodelling and other changes in the epigenome with consequent increased metastasis and invasiveness of the cancer (Gupta et al. 2010). The impact of *HOTAIR* on cancer development illustrates the importance of the non-coding network and the implications of dysregulation in disease, especially in cancer.

6.7 RNAs as Extracellular Signalling Molecules

Recent studies provide evidence for a role for RNAs in communication between cells (reviewed in Dinger et al. 2008). RNA has been shown to increase the permeability of endothelial cells in the blood–brain barrier through an interaction with vascular endothelial growth factor (Fischer et al. 2007). The roles of RNA as extracellular signaling molecules are less well understood than some of the other functions of non-coding RNAs but additional research is expected to uncover new pathways and mechanisms.

6.8 Pseudogenes and Non-coding RNA

A pseudogene is an imperfect copy of a functional gene, thought to arise during evolution as a result of retrotransposition or duplication. Previously dismissed as non-functional DNA, evidence shows that some pseudogenes are fully transcribed, resulting in the production of natural antisense transcripts (NAT). NATs are involved in numerous vital cellular processes, including regulation of translation and stability, RNA export, alternative splicing, genomic imprinting, X inactivation, DNA methylation and modification of histones, and have also been shown to play roles in stress response and developmental processes (Su et al. 2010). NATs transcribed from pseudogenes have the potential to regulate sense transcripts arising from the functional parental gene through complementary binding, which has been shown in some cases to induce cleavage of the sense transcript (Wilusz et al. 2009). Studies have shown that pseudogenes can also regulate their parental gene by interacting with enhancers, and that pseudogene transcripts can act as decoys for miRNAs targeting the parental gene (Poliseno et al. 2010) (reviewed in Muro et al. 2011). It is estimated that up to 20 % of human pseudogenes are fully transcribed (Zheng et al. 2007). However, it is likely that pseudogenes also produce smaller non-coding RNAs that may regulate gene expression *in cis* or *in trans*. Transcription of pseudogenes often occurs in a tissue-specific manner, and the discovery that pseudogenes are capable of regulating tumour suppressors and oncogenes and are often deregulated during cancer progression, indicates they are important components of the non-coding RNA regulatory system (reviewed in Pink et al. 2011). The discovery that pseudogenes may function in the form of non-coding RNAs shows that previous assumptions about "non-functional" regions of the human genome should be challenged.

6.9 Non-coding Capacity in Primates

Non-coding capacity is increased in primates in comparison to other animals. A comparison of pseudogenes across 28 vertebrate genomes showed that ~80 % of processed pseudogenes are primate-specific, indicating that the rate of retrotransposition is increased in primates (Zheng et al. 2007). Non-coding capacity is especially increased in the brain, with non-coding RNA a major contributor to evolution of gene expression pathways (Babbitt et al. 2010; Qureshi and Mehler 2012). RNA editing, a process by which bases are modified post-transcriptionally, is also predominantly active in the brain and is enriched in humans (Mattick and Mehler 2008), increasing diversity of the transcriptome (Paz-Yaacov et al. 2010). RNA editing is important as it allows adaptation to environmental stressors and may provide the basis for long-term memory and evolution of cognition throughout an individual's lifetime (Mattick and Mehler 2008). RNA editing also occurs extensively in non-coding RNAs, again highlighting the importance of these transcripts in the brain. A comparative genomics study on differences in humans that are highly conserved among other vertebrates, identified 202 elements of significance, mostly in non-coding regions (Pollard et al. 2006). It is clear that non-coding RNAs are key players in regulation and genome control and increasing organism complexity.

In the past decade research on non-coding RNAs has rapidly progressed, with hundreds of publications covering all known aspects of non-coding RNA function and regulation. For further information readers are referred to reviews on various subtopics: intron evolution and function (Mattick 1994); the significance of non-coding RNAs in organism complexity and evolution (Mattick 2001, 2003, 2011b; Prasanth and Spector 2007); functions of non-coding RNAs (Amaral et al. 2008; Mattick et al. 2009b), including regulation of transcription (Morris 2009; Hawkins and Morris 2008), epigenetic processes (Mattick et al. 2009a; Morris 2009), structural roles (Wilusz et al. 2009); and response to environmental stimuli (Varki et al. 2008); small regulatory RNAs in mammals (Mattick and Makunin 2005); non-coding RNAs in the human brain and development (Mattick 2011a; Mehler and Mattick 2006) and in the nervous system (Mehler and Mattick 2006); and the involvement of non-coding RNAs in disease (Taft et al. 2009).

6.10 Competing Endogenous RNAs

Competing endogenous RNA (ceRNA) is a newly discovered mechanism by which RNA molecules can regulate expression of one another by competing for miRNAs. As mentioned previously, transcripts originating from pseudogenes have been found to regulate the expression of the corresponding gene (Poliseno et al. 2010). Salmena et al. (2011) proposed that this idea is not limited to pseudogene transcripts, but that all types of RNA transcripts can communicate with one another via matching miRNA response elements (MREs) (Salmena et al. 2011). This mechanism of communication between mRNAs adds a new level of complexity in which the expression of miRNAs is affected by the targets as well as vice versa, creating elaborate regulatory networks. The more shared MREs between mRNAs, the greater chance of communication and co-regulation (Salmena et al. 2011). ceRNA activity is influenced by the relative concentrations of the ceRNAs and their miRNAs in a given cell at a particular time, and also the binding capacity of the MREs.

The most well studied example of ceRNA regulation involves the *PTEN* tumour suppressor gene. The *PTEN*-associated pseudogene has been shown to act as a ceRNA to regulate *PTEN*, with multiple conserved MREs allowing effective cross-talk between the two transcripts (Poliseno et al. 2010). This was experimentally demonstrated by overexpression of the pseudogene 3′UTR that resulted in a significant increase in the levels of PTEN. Pseudogene transcripts are particularly suited as competing RNAs with the associated gene, because the high-sequence conservation implies that they contain the same MREs. In addition, a number of other protein-coding transcripts that regulate PTEN in a miRNA-dependent manner have been identified, such as *SERINC1, VAPA*, and *CNOT6L* (Tay et al. 2011). Studying ceRNA pathways is likely to be a useful tool for gaining insight into the changes that come about during tumour growth. Research using an *in vivo* mouse model of melanoma confirmed the ceRNA relationships discovered by Tay et al. (2011) and validated the contribution of the ceRNAs in tumour growth and development (Karreth et al. 2011).

Although mRNAs from protein-coding genes can act as ceRNAs, it has been suggested that non-coding RNAs are likely to be overrepresented as highly effective regulators as they may be specifically synthesized for the purpose of regulation and there is no interference from active translation (Salmena et al. 2011). A recent study identified a muscle-specific long non-coding RNA, *linc-MD1*, that plays an important role in muscle differentiation by acting as a ceRNA in mouse and human myoblasts (Cesana et al. 2011). It was found that *linc-MD1* functions as a decoy for a number of miRNAs prevalent in muscle that are known to regulate the expression of multiple mRNAs. Targets of particular interest were *MAML1* and *MEF2C* that are muscle-specific transcription factors involved in myogenesis. Data demonstrated that *linc-MD1* communicates with these transcription factors as a ceRNA to regulate their expression (Cesana et al. 2011). Interestingly, the levels of *linc-MD1* were significantly reduced in Duchenne muscular dystrophy cells, and the accumulation of muscle-specific markers MYOG and MHC were delayed, and it is possible that the disruption of this ceRNA pathway contributes to Duchenne muscular dystrophy pathology. The study also found that the activation of the *linc-MD1* promoter correlates with the formation of a DNA loop at the beginning of myogenesis (Cesana et al. 2011). This shows how a ceRNA pathway can be activated when required and provide specific and sensitive control of mRNA levels in the cell.

ceRNA reveals a potential non-coding function of mRNAs that is separate to the protein function, adding yet another layer of complexity to the genome. This also has implications for research in which a specific transcript is targeted for knockout or upregulation, as such an approach would disrupt any ceRNA pathways involving that mRNA.

6.11 Chimeric RNAs

It is clear that the discovery of the non-coding regulatory network blurs the definition of a gene and suggests a more fluid concept in which a gene locus can produce multiple transcripts that interact to produce different protein isoforms at the required expression level. The discovery of chimeric RNAs, which are mRNAs consisting of exons from 2 or more different annotated genes, further distorts the concept of a gene and has the potential to exponentially increase the complexity of the genome. Although some of these transcripts arise because of DNA rearrangements such as retrotransposition, there are numerous transcripts composed of sequences originating from non-contiguous and non-linear parts of the genome, meaning they must have arisen via RNA-mediated events (Gingeras 2009). Characterisation of gene boundaries for 492 protein-coding genes on human chromosomes 21 and 22 using 5′ and 3′ RACE found that 42 % of genes produced detectable transcripts containing exons mapping to other annotated genes (Djebali et al. 2012). This study also found that connections are non-random and that connected genes had similar expression and close proximity in the nucleus.

Chimeric RNAs are generally expressed in a tissue specific manner, and protein expression from some chimeras has been detected, although it is unlikely that this is the main function of these transcripts (Frenkel-Morgenstern et al. 2012). Chimeric RNAs have also been shown to be overexpressed in various tumours (Zhou et al. 2012). These features indicate that chimeric RNAs are likely to be biologically significant, although their roles remain to be determined. The discovery and potentially common occurrence of chimeric RNAs has implications for the complexity of the genome, and further research into the function of these transcripts and the types of genes that form them is essential.

7 Conclusion

The non-coding regions of the genome, including the 5′ and 3′UTRs, introns and intergenic regions, are vital for the precise regulation of gene expression and have evidently expanded during the evolution of complex organisms. In addition, the recently discovered ceRNA pathway also implicates a non-coding function for protein coding mRNAs, and evidence of pervasive transcription throughout the genome suggests that RNA is the most prevalent and versatile component of the gene regulatory network. This aim of this review was to discuss all the different mechanisms by which non-coding DNA and RNA contribute to the local and global expression profiles, with the numerous mechanisms of control outlined here demonstrating that this regulatory system is highly complex and sensitive. Adding to this complexity, regulation often occurs in a tissue- and developmental-specific manner, exponentially increasing the variation of expression from the genome. A typical gene is mostly non-coding sequence, and accumulated evidence shows that these regions facilitate specific expression of gene isoforms, in specific quantities, and enable rapid response to changing conditions.

The clear correlation between the relative amount of non-coding sequence and the complexity of an organism demonstrates that it is the control networks that are the most important for evolution. This is logical when one considers the enormous variation that can result from a single gene, mediated by layers of regulatory components acting combinatorially to modulate gene expression. Complexity is increased by alternative mechanisms of gene processing, rather than the addition of more genes, as this allows an exponential rather than a linear increase in gene products. Humans have over 400 different cell types, including 145 types of neurons (Vickaryous and Hall 2006), all of which share the same DNA, with the exception of mature red blood cells and gametes. The differentiation of cell types has thus occurred through variation in the regulation of genes at all levels: from turning genes on or off, to subtle regulation arising from variation in non-coding RNA interactions. That the most significant changes in primates and humans in comparison to other organisms are found in the non-coding regions (King and Wilson 1975; Pollard et al. 2006) and the brain (Babbitt et al. 2010) is not surprising. A study exploring the nature of deletions of sequences in humans, that are otherwise highly conserved between chimpanzee and other mammals, found that the human-specific deletions fell almost exclusively in the non-coding regions, and were enriched near genes involved in neural function and steroid hormone signalling (McLean et al. 2011).

Non-coding RNAs are emerging as the most important, under-researched area of gene regulation and organism evolution. The ability of RNA to compactly store information and efficiently transmit that information within and between cells makes it an ideal medium of communication and regulation. Previously thought of as just an intermediate step for getting to the more "important" layer of protein expression, the RNA component of the genome is the largest and most versatile, involved in every level of gene expression. In order to appreciate and understand the complexity of regulation in the genome it will be essential to utilise new technologies to detect and characterise non-coding RNAs, investigate how these interact with other elements, and elucidate their function. An understanding of the factors and elements involved in the regulation of a particular gene is of paramount importance when designing molecular therapies or when attempting to modulate the expression of a gene.

References

Al-Ahmadi W, Al-Ghamdi M, Al-Haj L, Al-Saif M, Khabar KS (2009) Alternative polyadenylation variants of the RNA binding protein, HuR: abundance, role of AU-rich elements and auto-Regulation. Nucleic Acids Res 37:3612–3624

Amaral PP, Dinger ME, Mercer TR, Mattick JS (2008) The eukaryotic genome as an RNA machine. Science 319:1787–1789

Anastasi G, Cutroneo G, Santoro G, Arco A, Rizzo G, Bramanti P, Rinaldi C, Sidoti A, Amato A, Favaloro A (2008) Costameric proteins in human skeletal muscle during muscular inactivity. J Anat 213:284–295

Andreassi C, Riccio A (2009) To localize or not to localize: mRNA fate is in 3′UTR ends. Trends Cell Biol 19:465–474

Ansari AZ (2009) Rioactivators: transcription activation by noncoding RNA. Crit Rev Biochem Mol Biol 44:50–61

Arce L, Yokoyama NN, Waterman ML (2006) Diversity of LEF/TCF action in development and disease. Oncogene 25:7492–7504

Babbitt CC, Fedrigo O, Pfefferle AD, Boyle AP, Horvath JE, Furey TS, Wray GA (2010) Both noncoding and protein-coding RNAs contribute to gene expression evolution in the primate brain. Genome Biol Evol 2:67–79

Baek D, Davis C, Ewing B, Gordon D, Green P (2007) Characterization and predictive discovery of evolutionarily conserved mammalian alternative promoters. Genome Res 17:145–155

Banerjee AK (1980) 5′-terminal cap structure in eucaryotic messenger ribonucleic acids. Microbiol Rev 44:175–205

Barreau C, Watrin T, Beverley Osborne H, Paillard L (2006) Protein expression is increased by a class III AU-rich element and tethered CUG-BP1. Biochem Biophys Res Commun 347:723–730

Bartel DP (2004) MicroRNAs: genomics, biogenesis, mechanism, and function. Cell 116:281–297

Beaudoin JD, Perreault JP (2010) 5′-UTR G-quadruplex structures acting as translational repressors. Nucleic Acids Res 38:7022–7036

Bellizzi D, Dato S, Cavalcante P, Covello G, di Cianni F, Passarino G, Rose G, de Benedictis G (2007) Characterization of a bidirectional promoter shared between two human genes related to aging: SIRT3 and PSMD13. Genomics 89:143–150

Bharti K, Liu W, Csermely T, Bertuzzi S, Arnheiter H (2008) Alternative promoter use in eye development: the complex role and regulation of the transcription factor MITF. Development 135:1169–1178

Bianchi M, Crinelli R, Giacomini E, Carloni E, Magnani M (2009) A potent enhancer element in the 5′-UTR intron is crucial for transcriptional regulation of the human ubiquitin C gene. Gene 448:88–101

Birney E, Stamatoyannopoulos JA, Dutta A, Guigo R, Gingeras TR, Margulies EH, Weng Z, Snyder M, Dermitzakis ET, Thurman RE, Kuehn MS, Taylor CM, Neph S, Koch CM, Asthana S, Malhotra A, Adzhubei I, Greenbaum JA, Andrews RM, Flicek P, Boyle PJ, Cao H, Carter NP, Clelland GK, Davis S, Day N, Dhami P, Dillon SC, Dorschner MO, Fiegler H, Giresi PG, Goldy J, Hawrylycz M, Haydock A, Humbert R, James KD, Johnson BE, Johnson EM, Frum TT, Rosenzweig ER, Karnani N, Lee K, Lefebvre GC, Navas PA, Neri F, Parker SC, Sabo PJ, Sandstrom R, Shafer A, Vetrie D, Weaver M, Wilcox S, Yu M, Collins FS, Dekker J, Lieb JD, Tullius TD, Crawford GE, Sunyaev S, Noble WS, Dunham I, Denoeud F, Reymond A, Kapranov P, Rozowsky J, Zheng D, Castelo R, Frankish A, Harrow J, Ghosh S, Sandelin A, Hofacker IL, Baertsch R, Keefe D, Dike S, Cheng J, Hirsch HA, Sekinger EA, Lagarde J, Abril JF, Shahab A, Flamm C, Fried C, Hackermuller J, Hertel J, Lindemeyer M, Missal K, Tanzer A, Washietl S, Korbel J, Emanuelsson O, Pedersen JS, Holroyd N, Taylor R, Swarbreck D, Matthews N, Dickson MC, Thomas DJ, Weirauch MT, Gilbert J et al (2007) Identification and analysis of functional elements in 1% of the human genome by the ENCODE pilot project. Nature 447:799–816

Blaschke RJ, Topfer C, Marchini A, Steinbeisser H, Janssen JW, Rappold GA (2003) Transcriptional and translational regulation of the Leri-Weill and Turner syndrome homeobox gene SHOX. J Biol Chem 278:47820–47826

Bonnal S, Schaeffer C, Creancier L, Clamens S, Moine H, Prats AC, Vagner S (2003) A single internal ribosome entry site containing a G quartet RNA structure drives fibroblast growth factor 2 gene expression at four alternative translation initiation codons. J biol chem 278:39330–39336

Bostjancic E, Zidar N, Stajer D, Glavac D (2010) MicroRNAs miR-1, miR-133a, miR-133b and miR-208 are dysregulated in human myocardial infarction. Cardiology 115:163–169

Bradnam KR, Korf I (2008) Longer first introns are a general property of eukaryotic gene structure. PLoS One 3:e3093

Burley SK, Roeder RG (1996) Biochemistry and structural biology of transcription factor IID (TFIID). Annu Rev Biochem 65:769–799

Calvo SE, Pagliarini DJ, Mootha VK (2009) Upstream open reading frames cause widespread reduction of protein expression and are polymorphic among humans. Proc Natl Acad Sci USA 106:7507–7512

Cao Q, Richter JD (2002) Dissolution of the maskin-eIF4E complex by cytoplasmic polyadenylation and poly(A)-binding protein controls cyclin B1 mRNA translation and oocyte maturation. EMBO J 21:3852–3862

Carmell MA, Xuan Z, Zhang MQ, Hannon GJ (2002) The Argonaute family: tentacles that reach into RNAi, developmental control, stem cell maintenance, and tumorigenesis. Genes Dev 16:2733–2742

Carninci P, Kasukawa T, Katayama S, Gough J, Frith MC, Maeda N, Oyama R, Ravasi T, Lenhard B, Wells C, Kodzius R, Shimokawa K, Bajic VB, Brenner SE, Batalov S, Forrest AR, Zavolan M, Davis MJ, Wilming LG, Aidinis V, Allen JE, Ambesi-Impiombato A, Apweiler R, Aturaliya RN, Bailey TL, Bansal M, Baxter L, Beisel KW, Bersano T, Bono H, Chalk AM, Chiu KP, Choudhary V, Christoffels A, Clutterbuck DR, Crowe ML, Dalla E,

Dalrymple BP, de Bono B (2005) The transcriptional landscape of the mammalian genome. Science 309:1559–1563

Carninci P, Sandelin A, Lenhard B, Katayama S, Shimokawa K, Ponjavic J, Semple CA, Taylor MS, Engstrom PG, Frith MC, Forrest AR, Alkema WB, Tan SL, Plessy C, Kodzius R, Ravasi T, Kasukawa T, Fukuda S, Kanamori-Katayama M, Kitazume Y, Kawaji H, Kai C, Nakamura M, Konno H, Nakano K, Mottagui-Tabar S, Arner P, Chesi A, Gustincich S, Persichetti F, Suzuki H, Grimmond SM, Wells CA, Orlando V, Wahlestedt C, Liu ET, Harbers M, Kawai J, Bajic VB, Hume DA, Hayashizaki Y (2006) Genome-wide analysis of mammalian promoter architecture and evolution. Nat Genet 38:626–635

Carter MS, Li S, Wilkinson MF (1996) A splicing-dependent regulatory mechanism that detects translation signals. EMBO J 15:5965–5975

Cenik C, Derti A, Mellor JC, Berriz GF, Roth FP (2010) Genome-wide functional analysis of human 5′ untranslated region introns. Genome Biol 11:R29

Cesana M, Cacchiarelli D, Legnini I, Santini T, Sthandier O, Chinappi M, Tramontano A, Bozzoni I (2011) A long noncoding RNA controls muscle differentiation by functioning as a competing endogenous RNA. Cell 147:358–369

Chamboredon S, Ciais D, Desroches-Castan A, Savi P, Bono F, Feige JJ, Cherradi N (2011) Hypoxia-inducible Factor-1{alpha} mRNA: a new target for destabilization by tristetraprolin in endothelial cells. Mol Biol Cell 22:3366–3378

Chang YF, Imam JS, Wilkinson MF (2007) The nonsense-mediated decay RNA surveillance pathway. Annu Rev Biochem 76:51–74

Chatterjee S, Pal JK (2009) Role of 5′- and 3′-untranslated regions of mRNAs in human diseases. Biol Cell 101:251–262

Chatterjee S, Min L, Karuturi RK, Lufkin T (2010) The role of post-transcriptional RNA processing and plasmid vector sequences on transient transgene expression in zebrafish. Transgenic Res 19:299–304

Chen J, Kastan MB (2010) 5′–3′-UTR interactions regulate p53 mRNA translation and provide a target for modulating p53 induction after DNA damage. Genes Dev 24:2146–2156

Chen CY, Shyu AB (1995) AU-rich elements: characterization and importance in mRNA degradation. Trends Biochem Sci 20:465–470

Chen JM, Ferec C, Cooper DN (2006) A systematic analysis of disease-associated variants in the 3′ regulatory regions of human protein-coding genes II: the importance of mRNA secondary structure in assessing the functionality of 3′ UTR variants. Hum Genet 120:301–333

Cheng J, Kapranov P, Drenkow J, Dike S, Brubaker S, Patel S, Long J, Stern D, Tammana H, Helt G, Sementchenko V, Piccolboni A, Bekiranov S, Bailey DK, Ganesh M, Ghosh S, Bell I, Gerhard DS, Gingeras TR (2005) Transcriptional maps of 10 human chromosomes at 5-nucleotide resolution. Science 308:1149–1154

Clark MB, Mattick JS (2011) Long noncoding RNAs in cell biology. Semin Cell Dev Biol 22:366–376

Cleynen I, Brants JR, Peeters K, Deckers R, Debiec-Rychter M, Sciot R, van de Ven WJ, Petit MM (2007) HMGA2 regulates transcription of the Imp2 gene via an intronic regulatory element in cooperation with nuclear factor-kappaB. Mol Cancer Res 5:363–372

Cobbold LC, Spriggs KA, Haines SJ, Dobbyn HC, Hayes C, de Moor CH, Lilley KS, Bushell M, Willis AE (2008) Identification of internal ribosome entry segment (IRES)-trans-acting factors for the Myc family of IRESs. Mol Cell Biol 28:40–49

Cooper SJ, Trinklein ND, Anton ED, Nguyen L, Myers RM (2006) Comprehensive analysis of transcriptional promoter structure and function in 1% of the human genome. Genome Res 16:1–10

Cordes KR, Srivastava D (2009) MicroRNA regulation of cardiovascular development. Circ Res 104:724–732

Davuluri RV, Suzuki Y, Sugano S, Plass C, Huang TH (2008) The functional consequences of alternative promoter use in mammalian genomes. Trends Genet 24:167–177

Deaton AM, Bird A (2011) CpG islands and the regulation of transcription. Genes Dev 25:1010–1022

Dejosez M, Levine SS, Frampton GM, Whyte WA, Stratton SA, Barton MC, Gunaratne PH, Young RA, Zwaka TP (2010) Ronin/Hcf-1 binds to a hyperconserved enhancer element and regulates genes involved in the growth of embryonic stem cells. Genes Dev 24:1479–1484

Dhami P, Saffrey P, Bruce AW, Dillon SC, Chiang K, Bonhoure N, Koch CM, Bye J, James K, Foad NS, Ellis P, Watkins NA, Ouwehand WH, Langford C, Andrews RM, Dunham I, Vetrie D (2010) Complex exon-intron marking by histone modifications is not determined solely by nucleosome distribution. PLoS One 5:e12339

Dickson AM, Wilusz J (2010) Polyadenylation: alternative lifestyles of the A-rich (and famous?). EMBO J 29:1473–1474

Dieci G, Preti M, Montanini B (2009) Eukaryotic snoRNAs: a paradigm for gene expression flexibility. Genomics 94:83–88

Dinger ME, Mercer TR, Mattick JS (2008) RNAs as extracellular signaling molecules. J Mol Endocrinol 40:151–159

Djebali S, Lagarde J, Kapranov P, Lacroix V, Borel C, Mudge JM, Howald C, Foissac S, Ucla C, Chrast J, Ribeca P, Martin D, Murray RR, Yang X, Ghamsari L, Lin C, Bell I, Dumais E, Drenkow J, Tress ML, Gelpi JL, Orozco M, Valencia A, van Berkum NL, Lajoie BR, Vidal M, Stamatoyannopoulos J, Batut P, Dobin A, Harrow J, Hubbard T, Dekker J, Frankish A, Salehi-Ashtiani K, Reymond A, Antonarakis SE, Guigo R, Gingeras TR (2012) Evidence for transcript networks composed of chimeric RNAs in human cells. PLoS One 7:e28213

Djupedal I, Ekwall K (2009) Epigenetics: heterochromatin meets RNAi. Cell Res 19:282–295

Dmitriev SE, Andreev DE, Terenin IM, Olovnikov IA, Prassolov VS, Merrick WC, Shatsky IN (2007) Efficient translation initiation directed by the 900-nucleotide-long and GC-rich 5′ untranslated region of the human retrotransposon LINE-1 mRNA is strictly cap dependent rather than internal ribosome entry site mediated. Mol Cell Biol 27:4685–4697

Dmitriev SE, Andreev DE, Ad'Ianova ZV, Terenin IM, Shatskii IN (2009) Efficient cap-dependent in vitro and in vivo translation of mammalian mRNAs with long and highly structured 5′-untranslated regions. Mol Biol (Mosk) 43:119–125

Duan ZJ, Fang X, Rohde A, Han H, Stamatoyannopoulos G, Li Q (2002) Developmental specificity of recruitment of TBP to the TATA box of the human gamma-globin gene. Proc Natl Acad Sci USA 99:5509–5514

Duffy DL, Montgomery GW, Chen W, Zhao ZZ, Le L, James MR, Hayward NK, Martin NG, Sturm RA (2007) A three-single-nucleotide polymorphism haplotype in intron 1 of OCA2 explains most human eye-color variation. Am J Hum Genet 80:241–252

Eberhardt W, Doller A, Akool El S, Pfeilschifter J (2007) Modulation of mRNA stability as a novel therapeutic approach. Pharmacol Ther 114:56–73

Eberle AB, Stalder L, Mathys H, Orozco RZ, Muhlemann O (2008) Posttranscriptional gene regulation by spatial rearrangement of the 3′ untranslated region. PLoS Biol 6:e92

Eddy J, Maizels N (2008) Conserved elements with potential to form polymorphic G-quadruplex structures in the first intron of human genes. Nucleic Acids Res 36:1321–1333

Eisenberg I, Alexander MS, Kunkel LM (2009) miRNAS in normal and diseased skeletal muscle. J Cell Mol Med 13:2–11

Elbashir SM, Harborth J, Lendeckel W, Yalcin A, Weber K, Tuschl T (2001) Duplexes of 21-nucleotide RNAs mediate RNA interference in cultured mammalian cells. Nature 411:494–498

Elkon R, Zlotorynski E, Zeller KI, Agami R (2010) Major role for mRNA stability in shaping the kinetics of gene induction. BMC Genomics 11:259

Elton TS, Sansom SE, Martin MM (2010) Trisomy-21 gene dosage over-expression of miRNAs results in the haploinsufficiency of specific target proteins. RNA Biol 7:540–547

Fabian MR, Sonenberg N, Filipowicz W (2010) Regulation of mRNA translation and stability by microRNAs. Annu Rev Biochem 79:351–379

Fablet M, Bueno M, Potrzebowski L, Kaessmann H (2009) Evolutionary origin and functions of retrogene introns. Mol Biol Evol 26:2147–2156

Faghihi MA, Wahlestedt C (2009) Regulatory roles of natural antisense transcripts. Nat Rev Mol Cell Biol 10:637–643

Fedorova L, Fedorov A (2003) Introns in gene evolution. Genetica 118:123–131

Fejes-Toth K, Sotirova V, Sachidanandam R, Assaf G, Hannon GJ, Kapranov P, Foissac S, Willingham AT, Duttagupta R, Dumais E, Gingeras TR (2009) Post-transcriptional processing generates a diversity of 5'-modified long and short RNAs. Nature 457:1028–1032

Fialcowitz EJ, Brewer BY, Keenan BP, Wilson GM (2005) A hairpin-like structure within an AU-rich mRNA-destabilizing element regulates trans-factor binding selectivity and mRNA decay kinetics. J Biol Chem 280:22406–22417

Fichtlscherer S, de Rosa S, Fox H, Schwietz T, Fischer A, Liebetrau C, Weber M, Hamm CW, Roxe T, Muller-Ardogan M, Bonauer A, Zeiher AM, Dimmeler S (2010) Circulating microRNAs in patients with coronary artery disease. Circ Res 107:677–684

Filbin ME, Kieft JS (2009) Toward a structural understanding of IRES RNA function. Curr Opin Struct Biol 19:267–276

Fischer S, Gerriets T, Wessels C, Walberer M, Kostin S, Stolz E, Zheleva K, Hocke A, Hippenstiel S, Preissner KT (2007) Extracellular RNA mediates endothelial-cell permeability via vascular endothelial growth factor. Blood 110:2457–2465

Frenkel-Morgenstern M, Lacroix V, Ezkurdia I, Levin Y, Gabashvili A, Prilusky J, del Pozo A, Tress M, Johnson R, Guigo R, Valencia A (2012) Chimeras taking shape: potential functions of proteins encoded by chimeric RNA transcripts. Genome Res 22:1231–1242

Fritsch C, Herrmann A, Nothnagel M, Szafranski K, Huse K, Schumann F, Schreiber S, Platzer M, Krawczak M, Hampe J, Brosch M (2012) Genome-wide search for novel human uORFs and N-terminal protein extensions using ribosomal footprinting. Genome Res 22:2208–2218

Fu Y, Sun Y, Li Y, Li J, Rao X, Chen C, Xu A (2011) Differential genome-wide profiling of tandem 3'UTRs among human breast cancer and normal cells by high-throughput sequencing. Genome Res 21:741–747

Fukuchi M, Tsuda M (2010) Involvement of the 3'-untranslated region of the brain-derived neurotrophic factor gene in activity-dependent mRNA stabilization. J Neurochem 115:1222–1233

Gallie DR (1991) The cap and poly(A) tail function synergistically to regulate mRNA translational efficiency. Genes Dev 5:2108–2116

Ganapathi M, Srivastava P, Das Sutar SK, Kumar K, Dasgupta D, Pal Singh G, Brahmachari V, Brahmachari SK (2005) Comparative analysis of chromatin landscape in regulatory regions of human housekeeping and tissue specific genes. BMC Bioinformatics 6:126

Gerin I, Clerbaux LA, Haumont O, Lanthier N, Das AK, Burant CF, Leclercq IA, Macdougald OA, Bommer GT (2010) Expression of miR-33 from an SREBP2 intron inhibits cholesterol export and fatty acid oxidation. J Biol Chem 285:33652–33661

Ghosh T, Soni K, Scaria V, Halimani M, Bhattacharjee C, Pillai B (2008) MicroRNA-mediated up-regulation of an alternatively polyadenylated variant of the mouse cytoplasmic {beta}-actin gene. Nucleic Acids Res 36:6318–6332

Gilbert WV (2010) Alternative ways to think about cellular internal ribosome entry. J Biol Chem 285:29033–29038

Gingeras TR (2009) Implications of chimaeric non-co-linear transcripts. Nature 461:206–211

Girard A, Sachidanandam R, Hannon GJ, Carmell MA (2006) A germline-specific class of small RNAs binds mammalian Piwi proteins. Nature 442:199–202

Gomez D, Guedin A, Mergny JL, Salles B, Riou JF, Teulade-Fichou MP, Calsou P (2010) A G-quadruplex structure within the 5'-UTR of TRF2 mRNA represses translation in human cells. Nucleic Acids Res 38:7187–7198

Goodrich JA, Tjian R (2010) Unexpected roles for core promoter recognition factors in cell-type-specific transcription and gene regulation. Nat Rev Genet 11:549–558

Goodyer CG, Zheng H, Hendy GN (2001a) Alu elements in human growth hormone receptor gene 5' untranslated region exons. J Mol Endocrinol 27:357–366

Goodyer CG, Zogopoulos G, Schwartzbauer G, Zheng H, Hendy GN, Menon RK (2001b) Organization and evolution of the human growth hormone receptor gene 5'-flanking region. Endocrinology 142:1923–1934

Gorgoni B, Gray NK (2004) The roles of cytoplasmic poly(A)-binding proteins in regulating gene expression: a developmental perspective. Brief Funct Genomic Proteomic 3:125–141

Gorgoni B, Richardson WA, Burgess HM, Anderson RC, Wilkie GS, Gautier P, Martins JP, Brook M, Sheets MD, Gray NK (2011) Poly(A)-binding proteins are functionally distinct and have essential roles during vertebrate development. Proc Natl Acad Sci USA 108:7844–7849

Goss DJ, Kleiman FE (2013) Poly(A) binding proteins: are they all created equal? *Wiley interdisciplinary reviews*. RNA 4:167–179

Gu S, Jin L, Zhang F, Sarnow P, Kay MA (2009) Biological basis for restriction of microRNA targets to the 3′ untranslated region in mammalian mRNAs. Nat Struct Mol Biol 16:144–150

Guil S, Soler M, Portela A, Carrere J, Fonalleras E, Gomez A, Villanueva A, Esteller M (2012) Intronic RNAs mediate EZH2 regulation of epigenetic targets. Nat Struct Mol Biol 19:664–670

Gupta RA, Shah N, Wang KC, Kim J, Horlings HM, Wong DJ, Tsai MC, Hung T, Argani P, Rinn JL, Wang Y, Brzoska P, Kong B, Li R, West RB, van de Vijver MJ, Sukumar S, Chang HY (2010) Long non-coding RNA HOTAIR reprograms chromatin state to promote cancer metastasis. Nature 464:1071–1076

Hamilton AJ, Baulcombe DC (1999) A species of small antisense RNA in posttranscriptional gene silencing in plants. Science 286:950–952

Hawkins PG, Morris KV (2008) RNA and transcriptional modulation of gene expression. Cell Cycle 7:602–607

He Y, Vogelstein B, Velculescu VE, Papadopoulos N, Kinzler KW (2008) The antisense transcriptomes of human cells. Science 322:1855–1857

Holcik M, Lefebvre C, Yeh C, Chow T, Korneluk RG (1999) A new internal-ribosome-entry-site motif potentiates XIAP-mediated cytoprotection. Nat Cell Biol 1:190–192

Huang Y, Shen XJ, Zou Q, Wang SP, Tang SM, Zhang GZ (2011) Biological functions of microRNAs: a review. J Physiol Biochem 67:129–139

Hughes TA (2006) Regulation of gene expression by alternative untranslated regions. Trends Genet 22:119–122

Illingworth RS, Gruenewald-Schneider U, Webb S, Kerr AR, James KD, Turner DJ, Smith C, Harrison DJ, Andrews R, Bird AP (2010) Orphan CpG islands identify numerous conserved promoters in the mammalian genome. PLoS Genet 6:e1001134

Ivanov IP, Loughran G, Atkins JF (2008) uORFs with unusual translational start codons autoregulate expression of eukaryotic ornithine decarboxylase homologs. Proc Natl Acad Sci USA 105:10079–10084

Jackson RJ, Hellen CU, Pestova TV (2010) The mechanism of eukaryotic translation initiation and principles of its regulation. Nat Rev Mol Cell Biol 11:113–127

Jeffares DC, Penkett CJ, Bahler J (2008) Rapidly regulated genes are intron poor. Trends Genet 24:375–378

Jeffries CD, Fried HM, Perkins DO (2009) Additional layers of gene regulatory complexity from recently discovered microRNA mechanisms. Int J Biochem Cell Biol 42:1236–1242

Ji Z, Tian B (2009) Reprogramming of 3′ untranslated regions of mRNAs by alternative polyadenylation in generation of pluripotent stem cells from different cell types. PLoS One 4:e8419

Ji Z, Lee JY, Pan Z, Jiang B, Tian B (2009) Progressive lengthening of 3′ untranslated regions of mRNAs by alternative polyadenylation during mouse embryonic development. Proc Natl Acad Sci USA 106:7028–7033

Juven-Gershon T, Hsu JY, Theisen JW, Kadonaga JT (2008) The RNA polymerase II core promoter - the gateway to transcription. Curr Opin Cell Biol 20:253–259

Kandasamy K, Joseph K, Subramaniam K, Raymond JR, Tholanikunnel BG (2005) Translational control of beta2-adrenergic receptor mRNA by T-cell-restricted intracellular antigen-related protein. J Biol Chem 280:1931–1943

Kapp LD, Lorsch JR (2004) The molecular mechanics of eukaryotic translation. Annu Rev Biochem 73:657–704

Karreth FA, Tay Y, Perna D, Ala U, Tan SM, Rust AG, Denicola G, Webster KA, Weiss D, Perez-Mancera PA, Krauthammer M, Halaban R, Provero P, Adams DJ, Tuveson DA, Pandolfi PP (2011) *In vivo* identification of tumor- suppressive PTEN ceRNAs in an oncogenic BRAF-induced mouse model of melanoma. Cell 147:382–395

Kedde M, Strasser MJ, Boldajipour B (2007) RNA-binding protein Dnd1 inhibits microRNA access to target mRNA. Cell 131:1273–1286

King MC, Wilson AC (1975) Evolution at two levels in humans and chimpanzees. Science 188:107–116

Klattenhoff C, Theurkauf W (2008) Biogenesis and germline functions of piRNAs. Development 135:3–9

Knapinska AM, Gratacos FM, Krause CD, Hernandez K, Jensen AG, Bradley JJ, Wu X, Pestka S, Brewer G (2011) Chaperone Hsp27 modulates AUF1 proteolysis and AU-rich element-mediated mRNA degradation. Mol Cell Biol 31:1419–1431

Kochetov AV, Ischenko IV, Vorobiev DG, Kel AE, Babenko VN, Kisselev LL, Kolchanov NA (1998) Eukaryotic mRNAs encoding abundant and scarce proteins are statistically dissimilar in many structural features. FEBS Lett 440:351–355

Komar AA, Hatzoglou M (2005) Internal ribosome entry sites in cellular mRNAs: mystery of their existence. J Biol Chem 280:23425–23428

Kozak M (1987) Effects of intercistronic length on the efficiency of reinitiation by eucaryotic ribosomes. Mol Cell Biol 7:3438–3445.

Kozak M (1989) The scanning model for translation: an update. J Cell Biol 108:229–241

Krol J, Loedige I, Filipowicz W (2010) The widespread regulation of microRNA biogenesis, function and decay. Nat Rev Genet 11:597–610

Krummheuer J, Johnson AT, Hauber I, Kammler S, Anderson JL, Hauber J, Purcell DF, Schaal H (2007) A minimal uORF within the HIV-1 vpu leader allows efficient translation initiation at the downstream env AUG. Virology 363:261–271

Kuhn U, Gundel M, Knoth A, Kerwitz Y, Rudel S, Wahle E (2009) Poly(A) tail length is controlled by the nuclear poly(A)-binding protein regulating the interaction between poly(A) polymerase and the cleavage and polyadenylation specificity factor. J Biol Chem 284:22803–22814

Kumari S, Bugaut A, Huppert JL, Balasubramanian S (2007) An RNA G-quadruplex in the 5′ UTR of the NRAS proto-oncogene modulates translation. Nat Chem Biol 3:218–221

Kumari S, Bugaut A, Balasubramanian S (2008) Position and stability are determining factors for translation repression by an RNA G-quadruplex-forming sequence within the 5′ UTR of the NRAS proto-oncogene. Biochemistry 47:12664–12669

Lajevic MD, Koduvayur SP, Caffrey V, Cohen RL, Chambers DA (2010) Thy-1 mRNA destabilization by norepinephrine a 3′ UTR cAMP responsive decay element and involves RNA binding proteins. Brain Behav Immun 24:1078–1088

Lee I, Ajay SS, Yook JI, Kim HS, Hong SH, Kim NH, Dhanasekaran SM, Chinnaiyan AM, Athey BD (2009) New class of microRNA targets containing simultaneous 5′-UTR and 3′-UTR interaction sites. Genome Res 19:1175–1183

Lee JE, Lee JY, Wilusz J, Tian B, Wilusz CJ (2010) Systematic analysis of cis-elements in unstable mRNAs demonstrates that CUGBP1 is a key regulator of mRNA decay in muscle cells. PLoS One 5:e11201

Levine M, Tjian R (2003) Transcription regulation and animal diversity. Nature 424:147–151

Lifton RP, Goldberg ML, Karp RW, Hogness DS (1978) The organization of the histone genes in Drosophila melanogaster: functional and evolutionary implications. Cold Spring Harb Symp Quant Biol 42(Pt 2):1047–1051

Lin H (2007) piRNAs in the germ line. Science 316:397

Lin JM, Collins PJ, Trinklein ND, Fu Y, Xi H, Myers RM, Weng Z (2007) Transcription factor binding and modified histones in human bidirectional promoters. Genome Res 17:818–827

Liu L, Dilworth D, Gao L, Monzon J, Summers A, Lassam N, Hogg D (1999) Mutation of the CDKN2A 5′ UTR creates an aberrant initiation codon and predisposes to melanoma. Nat Genet 21:128–132

Lopez De Silanes I, Quesada MP, Esteller M (2007) Aberrant regulation of messenger RNA 3′-untranslated region in human cancer. Cell Oncol 29:1–17

Louro R, Smirnova AS, Verjovski-Almeida S (2009) Long intronic noncoding RNA transcription: expression noise or expression choice? Genomics 93:291–298

Lukavsky PJ (2009) Structure and function of HCV IRES domains. Virus Res 139:166–171

Lyabin DN, Eliseeva IA, Skabkina OV, Ovchinnikov LP (2011) Interplay between Y-box-binding protein 1 (YB-1) and poly(A) binding protein (PABP) in specific regulation of YB-1 mRNA translation. RNA Biol 8:883–892

Macdonald ME, Ambrose CM, Duyao MP, Myers RH, Lin C, Srinidhi L, Barnes G, Taylor SA, James M, Groot N, Macfarlane H, Jenkins B, Anderson MA, Wexler NS, Gusella JF, Bates GP, Baxendale S, Hummerich H, Kirby S, North M, Youngman S, Mott R, Zehetner G, Sedlacek Z, Poustka A, Frischauf A-M, Lehrach H, Buckler AJ, Church D, Doucette-Stamm L, O'Donovan MC, Riba-Ramirez L, Shah M, Stanton VP, Strobel SA, Draths KM, Wales JL, Dervan P, Housman DE, Altherr M, Shiang R, Thompson L, Fielder T, Wasmuth JJ, Tagle D, Valdes J, Elmer L, Allard M, Castilla L, Swaroop M, Blanchard K, Collins FS, Snell R, Holloway T, Gillespie K, Datson N, Shaw D, Harper PS (1993) A novel gene containing a trinucleotide repeat that is expanded and unstable on Huntington's disease chromosomes. Cell 72:971–983

Mangus DA, Evans MC, Jacobson A (2003) Poly(A)-binding proteins: multifunctional scaffolds for the post-transcriptional control of gene expression. Genome Biol 4:223

Mantovani R (1998) A survey of 178 NF-Y binding CCAAT boxes. Nucleic Acids Res 26:1135–1143

Martinez E, Chiang CM, GE H, Roeder RG (1994) TATA-binding protein-associated factor(s) in TFIID function through the initiator to direct basal transcription from a TATA-less class II promoter. EMBO J 13:3115–3126.

Martinez E, Zhou Q, L'etoile ND, Oelgeschlager T, Berk AJ, Roeder RG (1995) Core promoter-specific function of a mutant transcription factor TFIID defective in TATA-box binding. Proc Natl Acad Sci USA 92:11864–11868.

Maston GA, Evans SK, Green MR (2006) Transcriptional regulatory elements in the human genome. Annu Rev Genomics Hum Genet 7:29–59

Mathers JC, Strathdee G, Relton CL (2010) Induction of epigenetic alterations by dietary and other environmental factors. Adv Genet 71:3–39

Mattick JS (1994) Introns: evolution and function. Curr Opin Genet Dev 4:823–831

Mattick JS (2001) Non-coding RNAs: the architects of eukaryotic complexity. EMBO Rep 2:986–991

Mattick JS (2003) Challenging the dogma: the hidden layer of non-protein-coding RNAs in complex organisms. Bioessays 25:930–939

Mattick JS (2004) RNA regulation: a new genetics? Nat Rev Genet 5:316–323

Mattick JS (2011a) The central role of RNA in human development and cognition. FEBS Lett 585:1600–1616

Mattick JS (2011b) The central role of RNA in the genetic programming of complex organisms. An Acad Bras Cienc 82:933–939

Mattick JS, Makunin IV (2005) Small regulatory RNAs in mammals. Hum Mol Genet 14(Spec No 1):R121–R132

Mattick JS, Mehler MF (2008) RNA editing, DNA recoding and the evolution of human cognition. Trends Neurosci 31:227–233

Mattick JS, Amaral PP, Dinger ME, Mercer TR, Mehler MF (2009a) RNA regulation of epigenetic processes. Bioessays 31:51–59

Mattick JS, Taft RJ, Faulkner GJ (2009b) A global view of genomic information–moving beyond the gene and the master regulator. Trends Genet 26:21–28

Mayr C, Bartel DP (2009) Widespread shortening of 3′UTRs by alternative cleavage and polyadenylation activates oncogenes in cancer cells. Cell 138:673–684

Mazumder B, Seshadri V, Fox PL (2003) Translational control by the 3′-UTR: the ends specify the means. Trends Biochem Sci 28:91–98

McClelland S, Shrivastava R, Medh JD (2009) Regulation of Translational Efficiency by Disparate 5′UTRs of PPARgamma Splice Variants. PPAR Res 2009:193413

McLean CY, Reno PL, Pollen AA, Bassan AI, Capellini TD, Guenther C, Indjeian VB, Lim X, Menke DB, Schaar BT, Wenger AM, Bejerano G, Kingsley DM (2011) Human-specific loss of regulatory DNA and the evolution of human-specific traits. Nature 471:216–219

Mehler MF, Mattick JS (2006) Non-coding RNAs in the nervous system. J Physiol 575:333–341

Meijer HA, Thomas AA (2002) Control of eukaryotic protein synthesis by upstream open reading frames in the 5′-untranslated region of an mRNA. Biochem J 367:1–11

Meisner NC, Hackermuller J, Uhl V, Aszodi A, Jaritz M, Auer M (2004) mRNA openers and closers: modulating AU-rich element-controlled mRNA stability by a molecular switch in mRNA secondary structure. ChemBioChem 5:1432–1447

Mendell JT, Sharifi NA, Meyers JL, Martinez-Murillo F, Dietz HC (2004) Nonsense surveillance regulates expression of diverse classes of mammalian transcripts and mutes genomic noise. Nat Genet 36:1073–1078

Mercer TR, Dinger ME, Mattick JS (2009) Long non-coding RNAs: insights into functions. Nat Rev Genet 10:155–159

Mercer TR, Wilhelm D, Dinger ME, Solda G, Korbie DJ, Glazov EA, Truong V, Schwenke M, Simons C, Matthaei KI, Saint R, Koopman P, Mattick JS (2010) Expression of distinct RNAs from 3′ untranslated regions. Nucleic Acids Res 39:2393–2403

Meyer S, Temme C, Wahle E (2004) Messenger RNA turnover in eukaryotes: pathways and enzymes. Crit Rev Biochem Mol Biol 39:197–216

Mignone F, Gissi C, Liuni S, Pesole G (2002) Untranslated regions of mRNAs. Genome Biol 3:reviews0004

Mitchell SF, Walker SE, Algire MA, Park EH, Hinnebusch AG, Lorsch JR (2010) The 5′-7-methylguanosine cap on eukaryotic mRNAs serves both to stimulate canonical translation initiation and to block an alternative pathway. Mol Cell 39:950–962

Modarresi F, Faghihi MA, Lopez-Toledano MA, Fatemi RP, Magistri M, Brothers SP, van der Brug MP, Wahlestedt C (2012) Inhibition of natural antisense transcripts in vivo results in gene-specific transcriptional upregulation. Nat Biotechnol 30:453–459

Morris KV (2009) RNA-directed transcriptional gene silencing and activation in human cells. Oligonucleotides 19:299–306

Morris DR, Geballe AP (2000) Upstream open reading frames as regulators of mRNA translation. Mol Cell Biol 20:8635–8642

Mortensen RD, Serra M, Steitz JA, Vasudevan S (2011) Posttranscriptional activation of gene expression in Xenopus laevis oocytes by microRNA-protein complexes (microRNPs). Proc Natl Acad Sci USA 108:8281–8286

Munzarova V, Panek J, Gunisova S, Danyi I, Szamecz B, Valasek LS (2011) Translation reinitiation relies on the interaction between eIF3a/TIF32 and progressively folded cis-acting mRNA elements preceding short uORFs. PLoS Genet 7:e1002137

Muro EM, Mah N, Andrade-Navarro MA (2011) Functional evidence of post-transcriptional regulation by pseudogenes. Biochimie 93:1916–1921

Nahkuri S, Taft RJ, Mattick JS (2009) Nucleosomes are preferentially positioned at exons in somatic and sperm cells. Cell Cycle 8:3420–3424

Nakaya HI, Amaral PP, Louro R, Lopes A, Fachel AA, Moreira YB, El-Jundi TA, da Silva AM, Reis EM, Verjovski-Almeida S (2007) Genome mapping and expression analyses of human intronic noncoding RNAs reveal tissue-specific patterns and enrichment in genes related to regulation of transcription. Genome Biol 8:R43

Nolis IK, McKay DJ, Mantouvalou E, Lomvardas S, Merika M, Thanos D (2009) Transcription factors mediate long-range enhancer-promoter interactions. Proc Natl Acad Sci USA 106:20222–20227

O'Connell RM, Taganov KD, Boldin MP, Cheng G, Baltimore D (2007) MicroRNA-155 is induced during the macrophage inflammatory response. Proc Natl Acad Sci USA 104:1604–1609

Okada M, Nakajima K, Yaoita Y (2012) Translational regulation by the 5'-UTR of thyroid hormone receptor alpha mRNA. J Biochem 151:519–531

Okamura K, Lai EC (2008) Endogenous small interfering RNAs in animals. Nat Rev Mol Cell Biol 9:673–678

Ong CT, Corces VG (2011) Enhancer function: new insights into the regulation of tissue-specific gene expression. Nat Rev Genet 12:283–293

Ott CJ, Blackledge NP, Kerschner JL, Leir SH, Crawford GE, Cotton CU, Harris A (2009) Intronic enhancers coordinate epithelial-specific looping of the active CFTR locus. Proc Natl Acad Sci USA 106:19934–19939

Oyama M, Itagaki C, Hata H, Suzuki Y, Izumi T, Natsume T, Isobe T, Sugano S (2004) Analysis of small human proteins reveals the translation of upstream open reading frames of mRNAs. Genome Res 14:2048–2052

Oyama M, Kozuka-Hata H, Suzuki Y, Semba K, Yamamoto T, Sugano S (2007) Diversity of translation start sites may define increased complexity of the human short ORFeome. Mol Cell Proteomics 6:1000–1006

Paik JH, Jang JY, Jeon YK, Kim WY, Kim TM, Heo DS, Kim CW (2011) MicroRNA-146a Downregulates NF{kappa}B Activity via Targeting TRAF6 and Functions as a Tumor Suppressor Having Strong Prognostic Implications in NK/T Cell Lymphoma. Clin Cancer Res 17:4761–4771

Patikoglou GA, Kim JL, Sun L, Yang SH, Kodadek T, Burley SK (1999) TATA element recognition by the TATA box-binding protein has been conserved throughout evolution. Gene Dev 13:3217–3230.

Paz-Yaacov N, Levanon EY, Nevo E, Kinar Y, Harmelin A, Jacob-Hirsch J, Amariglio N, Eisenberg E, Rechavi G (2010) Adenosine-to-inosine RNA editing shapes transcriptome diversity in primates. Proc Natl Acad Sci USA 107:12174–12179

Peabody DS (1989) Translation initiation at non-AUG triplets in mammalian cells. J Biol Chem 264:5031–5035

Pelham HR, Jackson RJ (1976) An efficient mRNA-dependent translation system from reticulocyte lysates. Eur J Biochem 67:247–256

Pesole G, Mignone F, Gissi C, Grillo G, Licciulli F, Liuni S (2001) Structural and functional features of eukaryotic mRNA untranslated regions. Gene 276:73–81

Pickering BM, Willis AE (2005) The implications of structured 5' untranslated regions on translation and disease. Semin Cell Dev Biol 16:39–47

Pink RC, Wicks K, Caley DP, Punch EK, Jacobs L, Carter DR (2011) Pseudogenes: pseudo-functional or key regulators in health and disease? RNA 17:792–798

Poliseno L, Salmena L, Zhang J, Carver B, Haveman WJ, Pandolfi PP (2010) A coding-independent function of gene and pseudogene mRNAs regulates tumour biology. Nature 465:1033–1038

Pollard KS, Salama SR, King B, Kern AD, Dreszer T, Katzman S, Siepel A, Pedersen JS, Bejerano G, Baertsch R, Rosenbloom KR, Kent J, Haussler D (2006) Forces shaping the fastest evolving regions in the human genome. PLoS Genet 2:e168

Ponting CP, Oliver PL, Reik W (2009) Evolution and functions of long noncoding RNAs. Cell 136:629–641

Prabhakar S, Visel A, Akiyama JA, Shoukry M, Lewis KD, Holt A, Plajzer-Frick I, Morrison H, Fitzpatrick DR, Afzal V, Pennacchio LA, Rubin EM, Noonan JP (2008) Human-specific gain of function in a developmental enhancer. Science 321:1346–1350

Prasanth KV, Spector DL (2007) Eukaryotic regulatory RNAs: an answer to the 'genome complexity' conundrum. Genes Dev 21:11–42

Preker P, Nielsen J, Kammler S, Lykke-Andersen S, Christensen MS, Mapendano CK, Schierup MH, Jensen TH (2008) RNA exosome depletion reveals transcription upstream of active human promoters. Science 322:1851–1854

Preker P, Almvig K, Christensen MS, Valen E, Mapendano CK, Sandelin A, Jensen TH (2011) PROMoter uPstream Transcripts share characteristics with mRNAs and are produced upstream of all three major types of mammalian promoters. Nucleic Acids Res 39:7179–7193

Qureshi IA, Mehler MF (2012) Emerging roles of non-coding RNAs in brain evolution, development, plasticity and disease. Nat Rev Neurosci 13:528–541

Rapti A, Trangas T, Samiotaki M, Ioannidis P, Dimitriadis E, Meristoudis C, Veletza S, Courtis N (2010) The structure of the 5′-untranslated region of mammalian poly(A) polymerase-alpha mRNA suggests a mechanism of translational regulation. Mol Cell Biochem 340:91–96

Raveh-Amit H, Maissel A, Poller J, Marom L, Elroy-Stein O, Shapira M, Livneh E (2009) Translational control of protein kinase Ceta by two upstream open reading frames. Mol Cell Biol 29:6140–6148

Resch AM, Ogurtsov AY, Rogozin IB, Shabalina SA, Koonin EV (2009) Evolution of alternative and constitutive regions of mammalian 5′UTRs. BMC Genomics 10:162

Reznik B, Lykke-Andersen J (2010) Regulated and quality-control mRNA turnover pathways in eukaryotes. Biochem Soc Trans 38:1506–1510

Riley A, Jordan LE, Holcik M (2010) Distinct 5′UTRs regulate XIAP expression under normal growth conditions and during cellular stress. Nucleic Acids Res 38:4665–4674

Ringner M, Krogh M (2005) Folding free energies of 5′-UTRs impact post-transcriptional regulation on a genomic scale in yeast. PLoS Comput Biol 1:e72

Rodriguez-Trelles F, Tarrio R, Ayala FJ (2006) Origins and evolution of spliceosomal introns. Annu Rev Genet 40:47–76

Rose AB (2008) Intron-mediated regulation of gene expression. Curr Top Microbiol Immunol 326:277–290

Rotem-Dai N, Oberkovitz G, Abu-Ghanem S, Livneh E (2009) PKCeta confers protection against apoptosis by inhibiting the pro-apoptotic JNK activity in MCF-7 cells. Exp Cell Res 315:2616–2623

Saito T, Saetrom P (2010) MicroRNAs–targeting and target prediction. N Biotechnol 27:243–249

Salmena L, Poliseno L, Tay Y, Kats L, Pandolfi PP (2011) A ceRNA hypothesis: the Rosetta Stone of a hidden RNA language? Cell 146:353–358

Sandberg R, Neilson JR, Sarma A, Sharp PA, Burge CB (2008) Proliferating cells express mRNAs with shortened 3′ untranslated regions and fewer microRNA target sites. Science 320:1643–1647

Shah ZH, Toompuu M, Hakkinen T, Rovio AT, van Ravenswaay C, de Leenheer EM, Smith RJ, Cremers FP, Cremers CW, Jacobs HT (2001) Novel coding-region polymorphisms in mitochondrial seryl-tRNA synthetase (SARSM) and mitoribosomal protein S12 (RPMS12) genes in DFNA4 autosomal dominant deafness families. Hum Mutat 17:433–434

Shatsky IN, Dmitriev SE, Terenin IM, Andreev DE (2010) Cap- and IRES-independent scanning mechanism of translation initiation as an alternative to the concept of cellular IRESs. Mol Cells 30:285–293

Siepel A, Bejerano G, Pedersen JS, Hinrichs AS, Hou M, Rosenbloom K, Clawson H, Spieth J, Hillier LW, Richards S, Weinstock GM, Wilson RK, Gibbs RA, Kent WJ, Miller W, Haussler D (2005) Evolutionarily conserved elements in vertebrate, insect, worm, and yeast genomes. Genome Res 15:1034–1050

Smale ST, Kadonaga JT (2003) The RNA polymerase II core promoter. Annu Rev Biochem 72:449–479

Smith L (2008) Post-transcriptional regulation of gene expression by alternative 5′-untranslated regions in carcinogenesis. Biochem Soc Trans 36:708–711

Smith L, Brannan RA, Hanby AM, Shaaban AM, Verghese ET, Peter MB, Pollock S, Satheesha S, Szynkiewicz M, Speirs V, Hughes TA (2010a) Differential regulation of oestrogen receptor beta isoforms by 5′ untranslated regions in cancer. J Cell Mol Med, 14:2172–2184

Smith L, Coleman LJ, Cummings M, Satheesha S, Shaw SO, Speirs V, Hughes TA (2010b) Expression of oestrogen receptor beta isoforms is regulated by transcriptional and post-transcriptional mechanisms. Biochem J 429:283–290

Song B, Wang Y, Kudo K, Gavin EJ, Xi Y, Ju J (2008) miR-192 Regulates dihydrofolate reductase and cellular proliferation through the p53-microRNA circuit. Clin Cancer Res 14:8080–8086

Sotiropoulos A, Goujon L, Simonin G, Kelly PA, Postel-Vinay MC, Finidori J (1993) Evidence for generation of the growth hormone-binding protein through proteolysis of the growth hormone membrane receptor. Endocrinology 132:1863–1865

Southard JN, Barrett BA, Bikbulatova L, Ilkbahar Y, Wu K, Talamantes F (1995) Growth hormone (GH) receptor and GH-binding protein messenger ribonucleic acids with alternative 5'-untranslated regions are differentially expressed in mouse liver and placenta. Endocrinology 136:2913–2921

Stark A, Brennecke J, Bushati N, Russell RB, Cohen SM (2005) Animal MicroRNAs confer robustness to gene expression and have a significant impact on 3'UTR evolution. Cell 123:1133–1146

Stefanovic B, Brenner DA (2003) 5' stem-loop of collagen alpha 1(I) mRNA inhibits translation *in vitro* but is required for triple helical collagen synthesis *in vivo*. Journal Biol Chem 278:927–933

Sturm RA, Duffy DL, Zhao ZZ, Leite FP, Stark MS, Hayward NK, Martin NG, Montgomery GW (2008) A single SNP in an evolutionary conserved region within intron 86 of the HERC2 gene determines human blue-brown eye color. Am J Hum Genet 82:424–431

Su WY, Xiong H, Fang JY (2010) Natural antisense transcripts regulate gene expression in an epigenetic manner. Biochem Biophys Res Commun 396:177–181

Subramaniam K, Kandasamy K, Joseph K, Spicer EK, Tholanikunnel BG (2011) The 3'-untranslated region length and AU-rich RNA location modulate RNA-protein interaction and translational control of beta(2)-adrenergic receptor mRNA. Mol Cell Biochem 352:125–141

Suzuki Y, Holmes JB, Cerritelli SM, Sakhuja K, Minczuk M, Holt IJ, Crouch RJ (2010) An upstream open reading frame and the context of the two AUG codons affect the abundance of mitochondrial and nuclear RNase H1. Mol Cell Biol 30:5123–5134

Svitkin YV, Ovchinnikov LP, Dreyfuss G, Sonenberg N (1996) General RNA binding proteins render translation cap dependent. EMBO J 15:7147–7155

Taft RJ, Pang KC, Mercer TR, Dinger M, Mattick JS (2009) Non-coding RNAs: regulators of disease. J Pathol 220:126–139

Tan S, Guo J, Huang Q, Chen X, Li-Ling J, Li Q, Ma F (2007) Retained introns increase putative microRNA targets within 3'UTRs of human mRNA. FEBS Lett 581:1081–1086

Tanguay RL, Gallie DR (1996) Translational efficiency is regulated by the length of the 3' untranslated region. Mol Cell Biol 16:146–156

Tay Y, Kats L, Salmena L, Weiss D, Tan SM, Ala U, Karreth F, Poliseno L, Provero P, di Cunto F, Lieberman J, Rigoutsos I, Pandolfi PP (2011) Coding-independent regulation of the tumor suppressor PTEN by competing endogenous mRNAs. Cell 147:344–357

Torarinsson E, Sawera M, Havgaard JH, Fredholm M, Gorodkin J (2006) Thousands of corresponding human and mouse genomic regions unalignable in primary sequence contain common RNA structure. Genome Res 16:885–889

Torarinsson E, Yao Z, Wiklund ED, Bramsen JB, Hansen C, Kjems J, Tommerup N, Ruzzo WL, Gorodkin J (2008) Comparative genomics beyond sequence-based alignments: RNA structures in the ENCODE regions. Genome Res 18:242–251

Touriol C, Bornes S, Bonnal S, Audigier S, Prats H, Prats AC, Vagner S (2003) Generation of protein isoform diversity by alternative initiation of translation at non-AUG codons. Biol Cell 95:169–178

Varki A, Geschwind DH, Eichler EE (2008) Explaining human uniqueness: genome interactions with environment, behaviour and culture. Nat Rev Genet 9:749–763

Vasudevan S, Steitz JA (2007) AU-rich-element-mediated upregulation of translation by FXR1 and Argonaute 2. Cell 128:1105–1118

Vasudevan S, Tong Y, Steitz JA (2007) Switching from repression to activation: microRNAs can up-regulate translation. Science 318:1931–1934

Vickaryous MK, Hall BK (2006) Human cell type diversity, evolution, development, and classification with special reference to cells derived from the neural crest. Biol Rev Camb Philos Soc 81:425–455

Vlasova IA, Tahoe NM, Fan D, Larsson O, Rattenbacher B, Sternjohn JR, Vasdewani J, Karypis G, Reilly CS, Bitterman PB, Bohjanen PR (2008) Conserved GU-rich elements mediate mRNA decay by binding to CUG-binding protein 1. Mol Cell 29:263–270

Wahlestedt C (2006) Natural antisense and noncoding RNA transcripts as potential drug targets. Drug Discov Today 11:503–508

Wang ET, Sandberg R, Luo S, Khrebtukova I, Zhang L, Mayr C, Kingsmore SF, Schroth GP, Burge CB (2008) Alternative isoform regulation in human tissue transcriptomes. Nature 456:470–476

Wassarman DA, Steitz JA (1992) Interactions of small nuclear RNA's with precursor messenger RNA during *in vitro* splicing. Science 257:1918–1925

Watanabe T, Totoki Y, Toyoda A, Kaneda M, Kuramochi-Miyagawa S, Obata Y, Chiba H, Kohara Y, Kono T, Nakano T, Surani MA, Sakaki Y, Sasaki H (2008) Endogenous siRNAs from naturally formed dsRNAs regulate transcripts in mouse oocytes. Nature 453:539–543

Wen Y, Liu Y, Xu Y, Zhao Y, Hua R, Wang K, Sun M, Li Y, Yang S, Zhang XJ, Kruse R, Cichon S, Betz RC, Nothen MM, van Steensel MA, van Geel M, Steijlen PM, Hohl D, Huber M, Dunnill GS, Kennedy C, Messenger A, Munro CS, Terrinoni A, Hovnanian A, Bodemer C, de Prost Y, Paller AS, Irvine AD, Sinclair R, Green J, Shang D, Liu Q, Luo Y, Jiang L, Chen HD, Lo WH, McLean WH, He CD, Zhang X (2009) Loss-of-function mutations of an inhibitory upstream ORF in the human hairless transcript cause Marie Unna hereditary hypotrichosis. Nat Genet 41:228–233

Wethmar K, Smink JJ, Leutz A (2010) Upstream open reading frames: molecular switches in (patho)physiology. Bioessays 32:885–893

Williams AH, Liu N, van Rooij E, Olson EN (2009) MicroRNA control of muscle development and disease. Curr Opin Cell Biol 21:461–469

Wilusz JE, Sunwoo H, Spector DL (2009) Long noncoding RNAs: functional surprises from the RNA world. Genes Dev 23:1494–1504

Winter J, Kunath M, Roepcke S, Krause S, Schneider R, Schweiger S (2007) Alternative polyadenylation signals and promoters act in concert to control tissue-specific expression of the Opitz Syndrome gene MID1. BMC Mol Biol 8:105

Xie X, Lu J, Kulbokas EJ, Golub TR, Mootha V, Lindblad-Toh K, Lander ES, Kellis M (2005) Systematic discovery of regulatory motifs in human promoters and 3′UTRs by comparison of several mammals. Nature 434:338–345

Xu Z, Wei W, Gagneur J, Perocchi F, Clauder-Munster S, Camblong J, Guffanti E, Stutz F, Huber W, Steinmetz LM (2009) Bidirectional promoters generate pervasive transcription in yeast. Nature 457:1033–1037

Yang Z, Kaye DM (2009) Mechanistic insights into the link between a polymorphism of the 3′UTR of the SLC7A1 gene and hypertension. Hum Mutat 30:328–333

Yang C, Bolotin E, Jiang T, Sladek FM, Martinez E (2007) Prevalence of the initiator over the TATA box in human and yeast genes and identification of DNA motifs enriched in human TATA-less core promoters. Gene 389:52–65

Yekta S, Shih IH, Bartel DP (2004) MicroRNA-directed cleavage of HOXB8 mRNA. Science 304:594–596

Yokogawa T, Shimada N, Takeuchi N, Benkowski L, Suzuki T, Omori A, Ueda T, Nishikawa K, Spremulli LL, Watanabe K (2000) Characterization and tRNA recognition of mammalian mitochondrial seryl-tRNA synthetase. J Biol Chem 275:19913–19920

Zanotto E, Shah ZH, Jacobs HT (2007) The bidirectional promoter of two genes for the mitochondrial translational apparatus in mouse is regulated by an array of CCAAT boxes interacting with the transcription factor NF-Y. Nucleic Acids Res 35:664–677

Zhang R, Su B (2009) Small but influential: the role of microRNAs on gene regulatory network and 3′UTR evolution. J Genet Genomics 36:1–6

Zhang J, Tsaprailis G, Bowden GT (2008) Nucleolin stabilizes Bcl-X L messenger RNA in response to UVA irradiation. Cancer Res 68:1046–1054

Zhao Y, Srivastava D (2007) A developmental view of microRNA function. Trends Biochem Sci 32:189–197

Zhao Y, Ransom JF, Li A, Vedantham V, von Drehle M, Muth AN, Tsuchihashi T, McManus MT, Schwartz RJ, Srivastava D (2007) Dysregulation of cardiogenesis, cardiac conduction, and cell cycle in mice lacking miRNA-1-2. Cell 129:303–317

Zheng D, Frankish A, Baertsch R, Kapranov P, Reymond A, Choo SW, Lu Y, Denoeud F, Antonarakis SE, Snyder M, Ruan Y, Wei CL, Gingeras TR, Guigo R, Harrow J, Gerstein MB (2007) Pseudogenes in the ENCODE regions: consensus annotation, analysis of transcription, and evolution. Genome Res 17:839–851

Zhou J, Liao J, Zheng X, Shen H (2012) Chimeric RNAs as potential biomarkers for tumor diagnosis. BMB Rep 45:133–140